UNDERSTANDING PLANT PATHOLOGY

UNDERSTANDING PLANT PATHOLOGY

By

Dr. Shubhrata R. Mishra

Dept. of Botany
Vikram University
Ujjain (M.P.)
(India)

DISCOVERY PUBLISHING HOUSE PVT. LTD.
NEW DELHI-110 002

First Published-2010
Reprinted - 2014
ISBN 978-81-8356-514-1

Published by:
DISCOVERY PUBLISHING HOUSE PVT. LTD.
4831/24, Ansari Road, Prahlad Street,
Darya Ganj, New Delhi-110002 (India)
Phone: 23279245 • Fax: 91-11-23253475
E-mail: parul.wasan@gmail.com
info@discoverypublishinggroup.com
Website: www.discoverypublishinggroup.com

Printed at:
Dynamic Printers, Delhi

Preface

The present title "Understanding Plant Pathology" has been written for those students interested in careers in diverse fields of biological sciences. It provides a structured approach to learning by covering all the important topics in a uniform, systematic format. The book has been comprehensively designed incorporating recent advances in this fast moving field. It also provides accessible information on plant pathology in compact form for undergraduate students in biology and related life sciences. It is intelligible to the educated layman, though it deals with some complex ideas. It is an adequate text for all the requirements of students in this area. In addition, busy lecturers who require a quick reference compendium will find it useful, particularly for tutional planning. Simple, yet hopefully clear figures and tables are provided throughout the book.

The over-riding goal of this book, and indeed of the whole *Understanding series*, is to present the essential information concering plant pathology in a compact, readily accessible form which leads itself to student learning and revision. The convergence of various approaches has generated a rich panorama of detail, the significance of which we are still attempting to unraval. The present text has been written as an introduction to this rapidly growing field.

To make the work more comprehensive and informative, the author has consulted many authoritative books, research journals, abstracts, monographs etc., so there can be no claim to originality except in the manner of treatment.

The author expresses his thanks to his friends and colleagues whose continue inspirations have initiated him to bring out this book.

The author expresses his gratitude to Mr. Wasan and staff of M/s Discovery Publishing House Pvt. Ltd. for their whole hearted co-operation in the publication of this book.

In the mean time, the author will remain sincerely responsible for any shortcomings of the book and be grateful to the readers for their suggestions and constructive criticism for the continuous betterment of the book. He takes this opportunity to appeal to the readers to send their suggestions straightaway to his Publisher.

Author

CONTENTS

1

PLANT PATHOGENS

In 1892, the Russian virologist D. Iwanowski discovered horizontal transmission of viruses. He succeeded in infecting plants with the tobacco mosaic disease by applying a bacteria-free sap gained from infected tobacco leaves. The Dutchman M.W. Beijerinck confirmed this discovery some six years later. He filtered an extract of infected plants through a porcelain strainer not viable for bacteria, and found that the sap thus obtained displayed no loss of its infecting ability. The virus itself was not isolated till 1935 and not crystallized before 1937. In 1937, F.C. Bawden and N.W. Pirie (Rothamsted Experimental Station, England) detected that the preparations contained phosphate, an inherent part of the RNA molecule.

A number of TMV wild types differing, for example, in their host range or in the primary structure of their coat proteins exist. The classic strain that at the same time is the one that proliferates best on tobacco plants is called vulgare. A strain from tomatoes that proliferates equally well on tobacco plants is called dahlemense A third strain gained from plantain is known as Holmes ribgrass. The single strains differ visibly in the symptoms they cause in tobacco plants.

The amino acid sequences of the coat proteins of four different wild strains show that a certain sequence remains always unchanged. This sequence is assumed to be in direct contact with the RNA. The analysis of the coats' tertiary structures confirmed this assumption.

Characteristics of Plant Viruses

Detection: Small-require electron microscope for visualization. ELISA (Enzyme Linked Immuno Sorbent Assay) is often used to detect.

Morphology

1. *Rods*: Rigid rods - 5 × 300 nm Flexuous rods - 10-13 × 480-2000 nm
2. *Spheres*: Icosadodecahedron - 60 nm, Satellite - 17 nm Membrane encased - 100 nm, Bacillus-like, cylindrical rods 52-75 × 300-800 nm
3. *Monopartite*: All particles of the virus are the same size and contain the same nucleic acid.
4. *Multipartite*: Typically di- or tri- partite - virus population is made of two or three, respectively, sizes of particle with distinct nucleic acid strands encased in each capsid.

Composition and Structure

Each plant virus consists of at least one nucleic acid and one protein. Some viruses also have enzymes, lipids, and carbohydrates. The nucleic acid makes up 5 to 40 percent of the virus with the remaining 95 to 60 percent being protein. The proportion of nucleic acid to protein is greatest in spherical viruses. This simplest virus has a genome of approximately 1 to 3 × 10^6 daltons. This is in comparison with bacteria that have 1.5 × 10^9 daltons.

Composition and Structure of Viral Protein

The proteins of plant viruses are made of the same amino acids common to plant proteins. There are no unique amino acids distinctive of plant viruses. The protein shells of plant viruses are called capsids and are composed of repeating subunits. For any given virus the capsid proteins are identical and interchangeable though there is no universality of proteins between two different viruses.

In many cases a single protein composes the entire capsid, though more complex viral morphologies may require multiple protein types for their capsids. An individual virion is composed of a specific number of proteins unique to that virus particle. This is true irrespective of the morphological type of the virion. Plant viruses encode specific proteins that are involved in a variety of transport processes such as intra- and intercellular transport of viral genomes, long distance movement and transmission of virus particles between plants by vectors such as insects. It has extensive experience in studying the molecular biology of luteoviruses (Potato leafroll virus; PLRV) and umbraviruses (Groundnut rosette virus, Pea enation mosaic virus -2) that allows to use these viruses as experimental systems for analysing functions and biochemical properties of different "transport" proteins. Cell-to-cell and long-distance movement proteins encoded by umbraviruses have been identified. The methods used to explore how these proteins operate in plants include mutagenesis, immunocytochemical electron microscopy to investigate intracellular location, including nuclear and nucleolar targeting, biochemical studies such as of RNA-binding including the visualisation of RNA-protein complexes using atomic force microscopy. These approaches have the promise of providing powerful ways to investigate the modulation of plant transport processes.

Composition and Structure of Viral Nucleic Acid

The majority of plant viruses contain their genomes encoded as single strand RNA (ssRNA). However, double strand RNA (dsRNA), single strand DNA (ssDNA), and double strand DNA (dsDNA) plant viruses also occur. The nucleotide bases that make up plant virus RNA and DNA are the same as those typical to RNA and DNA of their host plants.

Satellite Viruses and Satellite RNAs

Some viruses require helper viruses to infect plants. These helper viruses are termed Satellite Viruses. "Satellite viruses are not related, or are only partially related, to the RNA of the virus; satellite RNAs may increase or decrease the severity of viral infections."

Amplicon-mediated RNA Silencing

An amplicon is a transgene that comprises a viral genome. The first to be produced was that of Tobacco mosaic virus (Yamaya et al., Mol. Gen. Genetics, 1988, 211, 520-525). Subsequently, using Potato virus X, Angell and Baulcombe (EMBO Journal, 1997, 16, 3675-3684) described how amplicons were capable of inducing RNA silencing.

Amplicons that consist of a full-length cDNA of the Potato leafroll virus (PLRV) genome or the same cDNA modified to contain DNA encoding green fluorescent protein (PLRV-GFP). PLRV and PLRV-GFP transgene-derived transcripts can infect protoplasts and thus because every cell of a transformed plant produces transcript, each should generate infectious RNA. However, replication of the transcribed amplicon was found to induce strong RNA silencing, as evidenced by the absence of virus symptoms, accumulation of only small amounts of PLRV-GFP RNA, low and non-uniform PLRV distribution and GFP expression, and accumulation of small 21-25 nt RNAs. These amplicons are being used to study the regulation of virus gene expression and the mechanisms of RNA silencing. This work is revealing the strategies used by viruses to combat host defence based on RNA silencing.

Biological Function of Viral Components

Coding

The function of the protein coat is not limited to encasing the viral genome. The outer surface of the protein is a major determinant in vector transmissibility. Barley Stripe Mosaic Virus contains a small carbohydrate component as a glycoprotein on the surface of its rod shaped capsid. This surface carbohydrate has been shown to be bound by lectins in barley seed and is thought to play a role in seed transmission of this virus.

Infectivity is strictly the province of the nucleic acid. In fact, under experimental conditions the nucleic acid, extracted away from its capsid, may be used to inoculate plants. Redundancy is not a feature of the viral genome. The amount on information necessary to code for the capsid protein is small (474 nucleotides for the 158 amino acids of Tobacco Mosaic Virus) relative to the total viral genome (6400 nucleotides); therefore it is obvious that the genome contains more information that that required to produce nucleocapsid proteins. The TMV genome codes for; nucleocapsid protein, two viral replicase (make more viral genome) enzymes, and a protein that facilitates movement through the plasmodesmata.

There is no direct evidence for coding for disease causing factors and it has been assumed that symptoms that are produced the result of disrupting host cellular process rather than specific viral metabolites.

There are three major groups of plant viruses that seem to be most important among new disease-causing agents worldwide and that could reasonably be termed "emerging" in terms of apparently being new viruses causing new and serious diseases. The most important of these are in the taxonomic families Potyviridae and Geminiviridae, the Tospovirus genus of the Bunyaviridae, and perhaps also pararetroviruses like Banana streak virus (BSV).

Potyviridae

The Potyviridae are the largest single taxonomic group of plant viruses, and the subject of more scientific papers per year than any other plant virus taxonomic group except perhaps the geminiviruses. It seems that potyviruses, and specifically aphid-transmitted viruses of the genus *Potyvirus*, are one of the most successful groups of plant pathogens in the world. The genus has been claimed to be almost as ancient as flowering plants, and has a worldwide distribution throughout higher plants. New potyviruses seem to appear with new crop introductions just about anywhere these occur. In South Africa, for example, *Passiflora edulis* (passionfruit) cultivation in the 1980s was severely limited by diseases, many of which appeared linked to a mixture of potyvirus-like agents. One of these viruses was characterized here in detail and found to be not the passionfruit woodiness virus (PWV) characterized in Australia and the Far East, but a distinct species now known to be a strain of cowpea aphid-borne mosaic virus (CABMV). It is very interesting that a similar disease in Brazil seems to be caused by a closely-related strain of the same virus, whereas woodiness disease in Australasia and Asia is caused by distinct species of viruses.

Another intriguing aspect of emerging potyviral plant diseases is the appearance of hitherto latent virus infections. A well-documented case investigated here was the appearance in cultivated specimens of the horticulturally important flowering bulbs Ornithogalum and Lachenalia species of at least two distinct potyviruses. One of these, Ornithogalum mosaic virus (OrMV), first described in South African-derived bulbs in Holland, was found in plant specimens collected in localities where the plants were endemic. However, it caused apparent or overt infections only when plants were kept in cultivation.

Geminiviridae

Viruses in the family Geminiviridae, and specifically the whitefly-transmitted begomoviruses, are presently associated with severe diseases in tomatoes the world over, with cotton disease in India and elsewhere, and with disease problems and food shortages due to disease in cassava in central Africa. Infections with Bean golden mosaic virus (BGMV) are apparently the single largest limiting factor to bean production in Central America.

Polston and Anderson (1997) covered the Caribbean basin and the Central American region in an excellent review of the emergence of whitefly-transmitted geminiviruses (WTGs) in tomatoes. They estimated that 20 - 100% of crop loss throughout these areas has been caused by epidemics of WTGs, ranging from Tomato golden mosaic (TGMV) to Tomato yellow leaf curl (TYLCV) to Potato yellow mosaic (PYMV). They listed 17 distinct viruses known to affect tomatoes in this region, and this is probably a conservative estimate, as more are constantly being found.

Much of the apparently emerging nature of the virus diseases is due to the worldwide spread of a new "silverleaf" or B biotype of the vector *Bemisia tabaci*, which is now being touted as the new species

Bemisia argentifolii. The new vector has a much wider range of preferences for feeding than older vector types, which has apparently resulted in the spread of viruses that normally infect only weed or endemic plant species into adjacent, previously untargeted crop species. Roye et al. (1997) surveyed geminiviruses in weeds (notably *Sida* and *Wissadula* species) in Jamaica and showed a population of viruses distinct from crop-infecting geminiviruses - and one perhaps poised to enter crops once a suitable vector biotype appears. Although the problem also exists in the developed world , and most notably in the southern United States and Europe, it can be dealt with, to at least some extent, by changing cultivation practices. The same option does not exist in the developing world, however, due to the expensive nature of the measures (screen houses, heavy spraying regimes). Thus, the inexorable spread of the vector into these areas will almost certainly mean heavy crop losses.

Begomovirus diseases also seem to exemplify a phenomenon first noted with potyviruses; unrelated or distinct virus species in the genus cause very similar disease syndromes in the same crop plants, usually in widely separated geographical areas but sometimes in the same small area. Thus, TYLCV-Israel, -Sardinia, and -Thailand are distinct species of virus despite causing the same disease; however, so are the several viruses causing cotton leaf curl disease in India and Pakistan, and the several distinct viruses causing tomato leaf curl in India. This is undoubtedly related to their worldwide distribution and also to their apparent antiquity. Geminivirus variation appears linked to geographic separation and host plant genetic divergence, as well as to continental drift. All these factors indicate that the youngest age for the family Geminiviridae is hundreds of millions of years and that of begomoviruses is at least tens of millions of years. This diversity and ubiquity guarantee that a wide variety of virus genotypes exists, probably as well-adapted and essentially symptomless infections in endemic species or weeds, and that any change in transmission characteristics of the vector will result in appearance of the virus in crop plants almost instantly.

Whereas begomoviruses have been making their presence felt as they spread out of endemic hosts into cultivated species, viruses in the genus *Mastrevirus* continue to emerge into crop species in Africa. Maize streak disease (MSD) was first described in 1901, so can hardly be considered as emerging. However, more recent findings indicate that a hitherto unsuspected and distinct group of Maize streak virus (MSV) strains cause disease in wheat and some grasses, rather than the disease being caused by the same closely-related group of virus isolates and strains as are found in maize (Rybicki et al., 1998; see also here). Sugarcane streak virus(es) (SSVs) also still seem to be present in Africa. Whereas the better characterized variants SSV-Natal and SSV-Mauritius have not been a problem in southern Africa for over 30 years, and especially not in commercially-grown sugarcane varieties, distinct mastrevirus species have recently been characterized from mildly diseased "traditional" material from Egypt.

A ray of hope in the struggle to combat TYLCV-like diseases is the fact that genetic engineering approaches using pathogen-derived resistance appear to be quite promising. Tomato plants transgenic for the TYLCV coat protein appear to be resistant to the virus (Kunik et al., 1994). Additionally, transgenic Nicotiana benthamiana plants expressing antisense RNA to the Rep gene were resistant to TYLCV infection (Bendahmane and Gronenborn, 1997). These developments, coupled with traditional and marker-assisted breeding techniques, may allow rapid dissemination of improved material to farmers.

Tospoviruses

The genus Tospovirus (family Bunyaviridae) is an extremely important group of plant viruses capable of infecting a large range of important crops. The type member, Tomato spotted wilt virus (TSWV), has one of the broadest host ranges among plant viruses. Tospoviruses are transmitted exclusively by thrips in a circulative propagative manner, meaning that the virus multiplies in the vector. TSWV was thought to be the only member of this genus until recently, and has had a serious impact on many crop species worldwide. It causes severe outbreaks in a large variety of crops grown in tropical and subtropical climate zones. Since 1990, other related but distinct viruses in this genus have been identified. The following members or tentative members are described: TSWV, Groundnut ringspot virus (GRSV), Tomato chlorotic spot virus (TCSV), Impatiens necrotic spot virus (INSV), Watermelon silver mottle virus (WSMV), and Groundnut bud necrosis virus (GBNV).

Owing to the worldwide spread of the Western flower thrip (Frankliniella occidentalis Pergande), the most efficient vector of tospoviruses, TSWV and some of the other tospoviruses are increasing being reported as causing problems in developing countries. These "new" tospoviruses often originate in developing countries, and some appear to occur only in these countries. For example, GRSV was first detected on peanuts during a survey of the viruses of this crop in South Africa, and isolate SA-05 now serves as the type isolate of GRSV. Thus far, this virus has been detected only on peanuts in South Africa, where it is relatively common, and on tomatoes in Brazil and Argentina. TCSV was first reported from Brazil, and GBNV was first detected in India.

Research on this important virus group has increased dramatically, and new tospoviruses are constantly being found. It could be that only the lack of resources in developing countries for the diagnosis of hard-to-identify viruses prevents even more reports of new tospoviruses.

Banana Streak Virus/Pararetroviruses

Banana streak virus (BSV) is a pararetrovirus - that is, a virus with circular dsDNA in virions which replicates via a longer-than-genome-length ssRNA(+) intermediate, via reverse transcription (see here). BSV may be considered an emerging problem in light of recent findings suggesting, first, that it occurs in an integrated form in Musa (see here); second, that it has been found in this form in all Musa germplasm tested thus far; third, that it is seed transmitted; and fourth, that the integrated virus can be activated to form the episomal form through stress. Stress can possibly include tissue culture propagation. Thus, it may be that initial BSV infections may largely be the result of plant stress, which means that BSV infections could occur anywhere, at any time, in the absence of any vectored introduction. It may be that genetic engineering in the form of a "knock-out" of the integrated viral genome can come to the rescue, at least of major commercial varieties.

Virus Infection of Plants

In economic terms, viruses are only of importance if it is likely that they will spread to crops during their commercial lifetime, which of course varies greatly between very short extremes in horticultural production and very long extremes in forestry. Some estimates put total worldwide damage due to plant viruses as high as US$ 6×10^{10} per year. Plant viruses face special problems initiating an infection. The outer surfaces of plants are composed of protective layers of waxes and pectin, but more significantly, a thick wall of cellulose overlying the cytoplasmic membrane surrounds each cell.

To date, no plant virus is known to use a specific cellular receptor of the type that animal and bacterial viruses use to attach to cells. Rather, plant viruses rely on a mechanical breach of the integrity of a cell wall to directly introduce a virus particle into a cell. This is achieved either by the vector associated with transmission of the virus or simply by mechanical damage to cells. After replication in an initial cell, the lack of receptors poses special problems for plant viruses in recruiting new cells to the infection.

Transmission of Plant Viruses

There are a number of routes by which plant viruses may be transmitted:

Seeds

These may transmit virus infection either due to external contamination of the seed with virus particles, or due to infection of the living tissues of the embryo. Transmission by this route leads to early outbreaks of disease in new crops, which are usually initially focal in distribution, but may subsequently be transmitted to the remainder of the crop by other mechanisms.

Vegetative propagation/grafting

These techniques are cheap and easy methods of plant propagation but provide the ideal opportunity for viruses to spread to new plants.

Vectors

Many different groups of living organisms can act as vectors and spread viruses from one plant to another:

- Bacteria (e.g. Agrobacterium tumefaciens - the Ti plasmid of this organism has been used experimentally to transmit virus genomes between plants)
- Fungi
- Nematodes
- Arthropods: Insects - aphids, leafhoppers, planthoppers, beetles, thrips, etc.
- Arachnids - mites

Mechanical

Mechanical transmission of viruses is the most widely used method for experimental infection of plants and is usually achieved by rubbing virus-containing preparations into the leaves, which in most plant species are particularly susceptible to infection. However, this is also an important natural method of transmission. Virus particles may contaminate soil for long periods and may be transmitted to the leaves of new host plants as wind-blown dust or as rain-splashed mud.

1. Transmission of plant viruses by insects is of particular agricultural importance. Extensive areas of monoculture and the inappropriate use of pesticides which kill natural predators can result in massive population booms of insects such as aphids.

2. Plant viruses rely on a mechanical breach of the integrity of a cell wall to directly introduce a virus particle into a cell. This is achieved either by the vector associated with transmission of the virus or simply by mechanical damage to cells.

Transfer by insect vectors is a particularly efficient means of virus transmission. In some instances, viruses are transmitted mechanically from one plant to the next by the vector and the insect is merely a means of distribution, flying or being carried on the wind for long distances (sometimes hundreds of miles). Insects which bite or suck plant tissues are, of course, the ideal means of transmitting viruses to new hosts. This is known as non-propagative transmission. However, in other cases (e.g. many plant rhabdoviruses) the virus may also infect and multiply in the tissues of the insect (propagative transmission) as well as those of host plants. In these cases, the vector serves as a means not only of distributing the virus, but also of amplifying the infection:

Group III geminivirus are transmitted by insect vectors (leafhoppers or whiteflies) and their genomes consist of two circular, single-stranded DNA molecules. These viruses cause a great deal of crop damage in plants such as tomatoes, beans, squash, cassava and cotton and their spread may be directly linked to the inadvertent world-wide dissemination of a particular biotype of the whitefly *Bemisia tabaci*. This vector is an indiscriminate feeder, encouraging rapid and efficient spread of viruses from indigenous plant species to neighbouring crops.

Multipartite Plant Viruses

1. Segmented virus genomes are those which are divided into two or more physically separate molecules of nucleic acid, all of which are then packaged into a single virus particle.
2. Although multipartite genomes also segmented, each genome segment is packaged into a separate virus particle.

Family	*Segments*
Geminivirus (group III) (single-stranded DNA)	Bipartite
Comovirus (single-stranded RNA)	Bipartite
Furovirus (single-stranded RNA)	Bipartite
Tobravirus (single-stranded RNA)	Bipartite
Partitiviridae (double-stranded RNA)	Bipartite
Bromoviridae (single-stranded RNA)	Tripartite
Hordeivirus (single-stranded RNA)	Tripartite

These discrete particles are structurally similar and may contain the same component proteins, but often differ in size depending on the length of the genome segment packaged. In one sense, multipartite genomes are, of course, segmented, but this is not the strict meaning of these terms as they are used here.

Genome segmentation reduces the probability of breakages due to shearing, thus increasing the total potential coding capacity of the genome. However, the disadvantage of this strategy is that all the individual genome segments must be packaged into each virus particle, or the virus will be defective as a result of loss of genetic information. Separating the genome segments into different particles (the multipartite strategy) removes the requirement for accurate sorting, but introduces a new problem in that all the discrete virus particles must be taken up by a single host cell to establish a productive infection. This is perhaps the reason multipartite viruses are only found in plants. Many of the sources of infection by plant viruses, such as inoculation by sap-sucking insects or after physical damage to tissues, result in a large inoculum of infectious virus particles, providing opportunities for infection of an initial cell by more than one particle.

Pathogenesis of Plant Virus Infections

Initially, most plant viruses multiply at the site of infection, giving rise to localized symptoms such as necrotic spots on the leaves. Subsequently, the virus may be distributed to all parts of the plant either by direct cell-to-cell spread or by the vascular system, resulting in a systemic infection involving the whole plant. However, the problem these viruses face in reinfection and recruitment of new cells is the same as they face initially - how to cross the barrier of the plant cell wall. "Plant cell walls necessarily contain channels called plasmodesmata which allow plant cells to communicate with each other and to pass metabolites between them. However, these channels are too small to allow the passage of virus particles or genomic nucleic acids."

Many (if not most) plant viruses have evolved specialized movement proteins which modify the plasmodesmata. One of the best known examples of this is the 30k protein of tobacco mosaic virus (TMV). This protein is expressed from a sub-genomic mRNA and its function is to modify plasmodesmata causing genomic RNA coated with 30k protein to be transported from the infected cell to neighbouring cells. Other viruses, such as cowpea mosaic virus (CPMV - Comovirus family) have a similar strategy but employ a different molecular mechanism. In CPMV, the 58/48k proteins form tubular structures allowing the passage of intact virus particles to pass from one cell to another.

Typically, virus infections of plants might result in effects such as growth retardation, distortion, mosaic patterning on the leaves, yellowing, wilting, etc. These macroscopic symptoms result from:

Plants might be seen as sitting targets for virus infection - unlike animals, they cannot run away. However, plants exhibit a range of responses to virus infections designed to minimize their effects. Initially, infection results in a 'hypersensitive response', manifested as the synthesis of a range of new proteins, the 'PR' (pathogenesis related) protein'. Although this system is poorly understood, at least some of these proteins have been characterized and have been shown to be proteases, which presumably destroy virus proteins, limiting the spread of the infection. There is some similarity here between this response and the production of interferons by animals. In addition, systemic resistance to virus infection is a naturally occurring phenomenon in some strains of plant, e.g. the tobacco N gene encodes a cytoplasmic protein with a nucleotide-binding site, which interferes with the TMV replicase. This is clearly a highly desirable characteristic and is highly prized by plant breeders, who try to spread this attribute to economically valuable crop strains.

1. *Necrosis* of cells, caused by direct damage due to virus replication
2. *Hypoplasia*, i.e. localized retarded growthfrequently leading to mosaicism (the appearance of thinner, yellow areas on the leaves)
3. *Hyperplasia*, which is excessive cell division or the growth of abnormally large cells, resulting in the production of swollen or distorted areas of the plant.

There are probably many different mechanisms involved in systemic resistance, but in general terms there is a tendency towards increased local necrosis as substances such as proteases and peroxidases are produced by the plant to destroy the virus and to prevent its spread and subsequent systemic disease. An example of this is the N gene, which when present in plants causes TMV to produce a localized, necrotic infection rather than the systemic mosaic symptoms normally seen.

Virus-resistant plants have been created by the production of transgenic plants expressing recombinant virus proteins or nucleic acids, which interfere with virus replication without producing the pathogenic consequences of infection, e.g:

1. Virus coat proteins: these have a variety of complex effects, including: (i) inhibition of virus uncoating; (ii) interference of

expression of the virus at the level of RNA ("gene silencing" by 'untranslatable' RNAs)

2. Intact or partial virus replicases which interfere with genome replication
3. Antisense RNAs
4. Defective viral genomes
5. Satellite sequences
6. Catalytic RNA sequences (ribozymes)
7. Modified movement proteins

This is a very promising technology which offers the possibility of substantial increases in agricultural production without the use of expensive, toxic and ecologically damaging chemicals (fertilizers, herbicides, or pesticides), but is at present still in its infancy.

Another developing aspect of plant biotechnology is the construction of CVPs - Chimaeric Virus Particles - recompinant viruses such as:

1. Cowpea mosaic virus - an icosahedral member of the Comovirus family

   or

2. Potato X virus- a rod-shaped member of the Potexvirus family which express short antigenic peptides on their surface. Several grams of CVPs can be produced cheaply and easily per kilogram of plant leaves. This is a very promising vaccine technology.

Emergent Plant Viruses

Rarely, there seems to be an example of an emergent virus which has acquired extra genes and as a result of new genetic capacity, becomes capable of infecting new species. A possible example of this phenomenon is seen in tomato spotted wilt virus (TSWV). TSWV is a Bunyavirus with a very wide plant host range, infecting over 600 different species from 70 families. In recent decades, this virus has been a major agricultural pest in Asia, the Americas, Europe Africa. Its rapid spread has been due to dissemination of its insect vector (the thrip Frankinellia occidentalis) and diseased plant material. TSWV is the type species of the Tospovirus genus and has a similar morphology and genomic organization to the other Bunyaviruses. However, TSWV undergoes propagative transmission and it has been suggested that it may have acquired an extra gene in the M segment via recombination, either from a plant or from another plant virus. This new gene encodes a movement protein, conferring the capacity to infect plants and cause extensive damage.

ICTV Approved Plant Virus Families and Genera

This list is a rearranged list of a large database containing descriptions of all viruses at the Australian National University's Bioinformatics Facility developed and maintained at the Research School of Biological Sciences. The rearangement divided plant viruses into serveral categories according to the type of viral genomic nucleic acids. The original list of the taxonomic structure of virus families and genera has been compiled from:

The Classification and Nomenclature of Viruses "Sixth Report of the International ommittee on Taxonomy of Viruses" (1995) Eds. F.A. Murphy, C.M. Fauquet, M.A. Mayo, A.W. Jarvis, S.A. Ghabrial, M.D. Summers, G.P. Martelli, D.H.L. Bishop Archives of Virology, Springer Verlag, Wien New York

Table 1.1. DNA Viruses

(A) Circular dsDNA Viruses

Family	*Genus*	*Type Species*
Caulimoviridae	Badnavirus	Commelina yellow mottle virus
	Caulimovirus	Cauliflower mosaic virus
	SbCMV-like viruses	Soybean chloroticmottle virus
	CsVMV-like viruses	Cassava vein mosaicvirus
	RTBV-like viruses	Rice tungro bacilliformvirus
	Petunia vein clearing-like viruses	Petunia vein clearing virus

(B) Circular ssDNA Viruses

Family	*Genus*	*Type Species*
Geminiviridae	Mastrevirus (Subgroup I Geminivirus)	Maize streak virus
	Curtovirus (Subgroup II Geminivirus)	Beet curly top
	Begomovirus (Subgroup III Geminivirus)	Bean golden mosaic virus

Throughout this document, each type species name is preceded by the name of the order, family and genus it belongs to. Names of orders, families and genera are in italic script, if they are approved by the ICTV. Taxa names in quotes (and not in italic script) indicate that this taxon has not an ICTV international approved name. Species (vernacular) names are given in regular script. Viruses with no formal assignment to genus or family are indicated.

Table 1.2. RNA Viruses

(A) ssRNA Viruses

Family	*Genus*	*Type Species*
Bromoviridae	Alfamovirus	Alfalfa mosaic virus
	Ilarvirus	Tobacco streak virus
	Bromovirus	Brome mosaic virus
	Cucumovirus	Cucumber mosaic virus
Closteroviridae	Closterovirus	Beet yellow virus
	Crinivirus	Lettuce infectious yellows virus
Comoviridae	Comovirus	Cowpea mosaic virus
	Fabavirus	Broad bean wilt virus
	Nepovirus	Tobacco ringspot virus
Potyviridae	Potyvirus	Potato virus Y
	Rymovirus	Ryegrass mosaic virus
	Bymovirus	Barley yellow mosaic virus
Sequiviridae	Sequivirus	Parsnip yellow fleck virus
	Waikavirus	Rice tungro spherical virus
Tombusviridae	Carmovirus	Carnation mottle virus
	Dianthovirus	Carnation ringspot virus
	Machlomovirus	Maize chlorotic mottle virus
	Necrovirus	Tobacco necrosis virus
	Tombusvirus	Tomato bushy stunt virus
Unassigned Genera of ssRNA viruses	Capillovirus	Apple stem grooving virus
	Carlavirus	Carnation latent virus
	Enamovirus	Pea enation mosaic virus
	Furovirus	Soil-borne wheat mosaic virus
	Hordeivirus	Barley stripe mosaic virus
	Idaeovirus	Raspberry bushydwarf virus
	Luteovirus	Barley yellow dwarf virus
	Marafivirus	Maize ravado fino virus
	Potexvirus	Potato virus X
	Sobemovirus	Southern bean mosaic virus
	Tenuivirus	Rice stripe virus
	Tobamovirus	Tobacco mosaic virus
	Tobravirus	Tobacco rattle virus
	Trichovirus	Apple chlorotic leaf spot virus
	Tymovirus	Turnip yellow mosaic virus
	Umbravirus	Carrot mottle virus

(B) Negative ssRNA Viruses: Order - Mononegavirales

Family	*Genus*	*Type Species*
Rhabdoviridae	Cytorhabdovirus	Lettuce necrotic yellow virus
	Nucleorhabdovirus	Potato yellow dwarf virus
Bunyaviridae	Tospovirus	Tomato spotted wilt virus

(C) dsRNA Viruses:

Family	*Genus*	*Type Species*
Partitiviridae	Alphacryptovirus	White clover cryptic virus 1
	Betacryptovirus	White clover cryptic virus 2
Reoviridae	Fijivirus	Fiji disease virus
	Phytoreovirus	Wound tumor virus
	Oryzavirus	Rice ragged stunt virus

Use of Phylogenetic Analysis in the Classification of Plant Viruses

The classification of plant viruses has traditionally been a conservative field of endeavour, with plant virologists resolutely resisting the introduction of Latinate nomenclature and Linnaean classification schemes. Plant viruses have also until recently been classified in only two taxa - the "virus" (more or less equivalent ot animal virus species) and the "group" (equivalent to genus).

Table 1.3. Unassigned Viruses.

Genome	*Species*
ssDNA	Banana bunchy top virus
	Coconut foliar decay virus
	Subterranean clover stunt virus
ds DNA	Cucumber vein yellowing virus
ds RNA	Tobacco stunt virus
ss RNA	Garlic viruses A,B,C,D
	Grapevine fleck virus
	Maize white line mosaic virus
	Olive latent virus 2
	Ourmia melon virus
	Pelargonium zonate spot virus

Recently, however, there has come a move to introduce higher-level taxa such as the "supergroup" or family into plant virus classification schemes, in line with animal and human virus classification. There has also been an increasing use of genome sequence data for establishing relationships between viruses: some of these analyses have been especially illuminating in that they have revealed distant relationships between several plant virus groups and some animal virus families.

Table 1.4. Satellites and Viroids.

Type	*Genome*	*Subgroup*	*Type species*
Satellites	ssRNA Satellite RNA	Subgroup 2A Subgroup 2B Type mRNA Satellites Subgroup 3C Type linear RNA Satellites Subgroup 4D Type circular RNA Satellites	Tobacco necrosis satellite
Viroids			Potato spindle tuber viroid

Two notable examples are the relationships between picornaviruses, comoviruses and potyviruses; and between caulimoviruses, retroviruses and hepadnaviruses.

Such sequence comparisons have largely made use of strongly-conserved predicted protein sequences that are related to replication; the inference of the analyses is that the replicases of individual viruses within these separate groups of viruses have diverged evolutionarily from a common source.

Similar sequence comparisons have been used to estimate evolutionary distances between, and make taxonomic proposals for, many different organisms, ranging from bacteria to plants to hominoids. A common approach in these studies has been the construction of phylogenetic trees for the organisms in question: these have been used to clarify lines of evolutionary descent of the organisms, and illustrate relationships with taxonomic significance. It is interesting that very few such analyses have been performed for viruses in general, and plant viruses in particular - probably both because of a general reluctance to speculate upon virus evolutionary processes, and because of a lack of sequence data. The analyses that have been done have also concentrated mainly on distance data - that is, pairwise similarity or difference estimates - for use in phenetic comparisons, rather than using the cladistic techniques which have become popular in animal and plant systematics recently.

Data Sets Used for Relationship Dendrograms and Phylo-genetic Analyses

Use of serological data

Some of the first quantitative estimates of "difference" between plant viruses were based on serological data. Serology has long been used for grouping plant viruses. However, the concept of quantitating serological differences is fairly recent. In 1970 van Regenmortel and von Wechmar introduced the concept of the serological differentiation index or SDI: this is the serological cross-reactivity between two viruses expressed as the number of twofold dilution steps separating homologous and heterologous titers.

Formerly such investigations were performed using precipitin techniques; recently, however, ELISA techniques have been used for SDI determinations: Jaegle and van Regenmortel demonstrated the general use of the technique, and Dekker et al. (1988) and Pinner et al. (1992) put it to use in determining relationships among maize streak virus (MSV) and related geminiviruses of cereals and grasses.

SDI data have been used for taxonomic purposes, in that "clusters" of related viruses can be differentiated from less-related species. It has also been used to construct relationship dendrograms: Koenig (1976) presented a comprehensive analysis of the serology of 13 tymoviruses, which included a novel "loop structure".

In fact, one can take a table of reciprocal SDI values for a group of viruses and analyse it directly for construction of a dendrogram. The SDI data is conceptually similar to the pairwise DNA hybridisation data that has often been used for construction of phylogenetic trees); thus, one may also root the dendrogram to make it into a tree, and speculate on the phylogeny of virus coat proteins from serological data.

Coat protein composition

Some virologists have shown that the relatedness of tobamovirus coat proteins as assessed from amino acid sequences agrees closely with that assessed from their amino acid compositions. These latter data to construct relationship dendrograms for many tobamoviruses: these dendrograms may also be regarded as phylogenetic trees, as they are rooted, and reflect a line of evolutionary descent. These may productively be used for re-classification of certain of the viruses, and as confirmation of the classification of several others: for example, closely-clustered viruses could be regarded as strains, and distinct clusters as groups of strains of different viruses.

That the approach may be comfortably used to define a taxonomic group - or genus - is shown by Gibbs (198C), where a tree that includes the furovirus beet necrotic yellow vein virus (BNYVV) clearly separates this from definitive tobamoviruses.

Fauquet et al. (1985) have also claimed that the amino acid compositions of most virus coat proteins fell into groups closely corresponding to accepted plant virus taxonomic groups: thus determination of the composition of a virus coat protein may well be a most useful exercise from a taxonomic point of view, as such data appears to define the virus taxonomic group. However, it is still an analytical undertaking of some magnitude, and difficult for most laboratories.

Limited RNAse digestion and oligonucleotide mapping

For many years researchers have been mapping and typing ssRNA viruses of mammals (and of plants) by limited RNAse digestion of genomic RNA and oligonucleotide mapping by electrophoresis and chromatography. Such analyses are obviously possible for plant viruses, and could be as informative. They are only really usable at the level of "species" relationships, as distantly-related RNAs tend to have very different oligonucleotide maps.

Restriction fragment and map data

It is a relatively simple task - and certainly easier than exhaustive protein purification protocols - to purify double-stranded DNAs from plants, and to analyse them by restriction endonuclease digestion, electrophoresis, and hybridisation with labelled probes. The use of restriction endonuclease cleavage patterns and derived restriction maps of DNA sequences for phylogenetic analysis - particularly of mitochondrial DNAs - is well established; however, the methods have had little application in plant virology since most plant viruses have ssRNA genomes.

One group of viruses that have been studied by restriction mapping and restriction fragment pattern differences are the geminiviruses of maize and cereals.

It is possible to directly convert a series of restriction fragment patterns of different viral DNAs on an electropherogram into a digital data matrix and analyse this by cladistic techniques; it is also possible to construct difference tables and analyse them phenetically. Detailed restriction maps of related virus DNAs may be transformed into distance matrices or also treated as digital information for cladistic analysis.

With recent developments in cDNA synthesis technology, it is also possible to routinely map RNA viruses by restriction enzyme cleavage: for instance, DNA fragments obtained from human rhinoviruses by polymerase chain reaction (PCR) amplification of cDNA have been used for typing of the virus isolates by restriction endonuclease cleavage patterns; PCR has also been used for amplification and typing of viroids from cDNA. Restriction map data falls down, however, at demonstrating relationships beyond the level of about 30% sequence difference. Thus map comparisons may only be useful for sub-group level comparisons.

Nucleic acid and protein sequences

Recent developments in nucleic acid technology make it almost easier to obtain long stretches of RNA or DNA sequence, than to sequence or to determine the composition of the protein coded for by the sequence; however, the latter has also become easier (Shukla and Ward, 1989). Thus, comparisons of coat protein gene sequences - or of the predicted amino acid sequences coded for - are probably going to be far more widely used for comparison purposes than are serological data or amino acid composition data. The usefulness of coat protein sequence differences for phylogenetic analysis of the tobamoviruses has already been mentioned; it is worth noting that Gibbs (1976) has used such an analysis to estimate a divergence time of between 200 and 600 Myr for the definitive tobamoviruses. Both sorts of analysis - directly from sequence, and from pairwise sequence difference comparisons - lead to diagrams which can be used to group viruses.

Application of Phylogenetic Analyses to Plant Virus Classification

A convenient rule of thumb with the simpler viruses appears to be that coat protein gene sequences vary the most, followed by certain of the non-structural proteins such as movement proteins, followed by proteins associated with the replication machinery. This is nowhere more clear than among the plant virus groups linked by possession of a tripartite ssRNA genome and isometric or bacilliform particles: the bromo-, cucumo-, ilar- and alfalfa mosaic virus groups, which have been proposed as members of a plant virus family to be called the Bromoviridae.

Table 1.5. Classification of plant virus based on phylogenetic analysis.

Organism (by category)	*Strain*	*Common Name*	*Genome Size*	*Rationale or Significance*
Potyvirus	MDMV	Maize dwarf mosaic	10.4 kb	Largest single taxonomic group of plant viruses, transmission by aphids (nonpersistentmanner), whitefly or fungi, some seedborne, include leguminous, solanaceous and monocot hosts, global distribution.
	WSMV	Wheat streak mosaic		
Tritimovirus	PVY	Potato Y		
	BCMV	Bean common mosaic		
	SMV	Soybean mosaic		
Closteroviruses /Criniviruses	CTV	Citrus tristeza	17-20 kb/15kb	Viruses of woody and herbaceous hosts. large complex genomes of ssRNA, semi-persistent,
	ToCV	Tomato chlorosis		
	PMWV	Pineapple mealy bug		Associated with hemipteran vectors (aphid, mealy bug and whitefly).
Begomoviruses	SLCV	Diverse strains and species:	5.2 kb	Only circular ssDNA plant viruses, global, greater than expected variability, apparent rapid evolution through recombination and reassortment, poor association between
	BGMV	Bipartite New World or		
	AbMV	Old World monopartite		
Curtoviruses	PHV clades BCTV and	viruses of cultivated and wild species	2.8 kb	viruses of cultivated and weed hosts, several hemipteran vectors

Continue...

Organism (by category)	*Strain*	*Common Name*	*Genome Size*	*Rationale or Significance*
	related viruses			
Tospoviruses	TSWV	Tomato spotted wilt	~17 kb	Large, complex, ssRNA virus replicates in insect vector. Medical significance (Bunyaviridae).
Tombusvirus	TBSV CNV	Tomato bushy stunt Cucumber necrosis	4.8 kb	Highly stable, ssRNA viruses transmitted in persistent manner by fungal zoospores
Luteovirus/ Polerovirus	BYDV BWYV	Barley yellow dwarf Beet western yellows	5.5 kb	ssRNA viruses transmitted in persistent, circulative manner by aphids, worldwide distribution, infect monocots and dicots.
Cucumovirus	CMV stains	Cucumber mosaic virus	8.0 kb	One of the largest groups of plant viruses, have broad host ranges, but poorly understood biodiversity. Global distribution.
Rhabdovirus	SYVV	Sonchus yellow vein virus	13.0 kb	Large, complex, membrane bound RNA viruses, replicates in insect, medical significance through relationship to animal viruses.

These examples could be taken as leading inexorably to a proposal for the establishment of plant virus taxa right up to the "super-familial" level. However, plant virus genome relationships are not as simple as are the phylogenetic relationships of some of their constituent parts and any taxon higher than the familial level would be very hard to justify. For example, the Bromoviridae all have a similar genomic structure and organisation, and it could be confidently asserted that all components of all of them have a common source. Tobamoviruses also all have a similar genetic organisation. Their putative replicases are related to those of the Bromoviridae; however, there is no demonstrable relationship between any other genes of the two groups of viruses, and the genetic organisation is markedly different. Moreover, both sets of viruses share sequence similarity with the animal alphaviruses, in the family Togaviridae, which have even less similar genetic organisation.

Much the same can be said for the picorna-, como- and potyviruses. Some virologists have proposed that simple viruses have evolved by building up modules into genomes, and that different "supergroups" can be distinguished by their core modules, which consist of genes related to replication. Thus there is a picornavirus-like supergroup of viruses with similar polymerse/protease core modules and an alphavirus supergroup of viruses with similar "nucleotide-binding" protein/replicase cores.

With plant viruses, the term specificity (or host-specificity) has a very narrow meaning, since no plant virus as such exists. Instead, plant viruses can be grouped in a number of 'varieties'. The tobacco mosaic-virus (TMV), for example, multiplies within Nicotiana-species, several other solanaceous plants, and a few species of other plant families. The name of a virus is usually derived from the name of its main host plant. Although with viruses, the term species may not quite correspond to the way it is defined in biological systematics, it is perfectly reasonable and common to use it for viruses, too, since all viruses and viroids contain an original genome with a species-specific information. Its continuity over generations is guaranteed by replication in the host cells. The genetic information of viruses is either encoded by single-stranded RNA (most plant viruses), double-stranded RNA (wound tumor viruses), single-stranded DNA (gemini-viruses) or double-stranded DNA (cauliflower mosaic-virus: CaMV). Based on the shape of the virus particle, it is distinguished between rod-shaped and icosaedrical viruses with a capsid that seems almost spherical.

Plant viruses have no specific mechanism for entering the host cell. Cell wall and cuticle are difficult obstacles for them. Plant viruses depend therefore on injuries or on transmission via invertebrates (insects, nematodes, etc.). The animal transmitter does in some cases also act as an intermediate host. This means that some plant viruses are able to multiply within animal tissue.

Virus diseases of plants are relatively rare. Infection is scarcely strong enough to kill the plant. Monoculture favours spreading, and agricultural losses of profit caused by the potato-X, or the potato-Y viruses, for example, can be significant.

Of many viruses, numerous, considerably differing strains (wildtypes) have been isolated. Among the differences are host range and the degree of virulence. Moreover, viroids were detected. Viroids are small, circular RNA molecules that do not encode proteins themselves. Instead, virions interfere with the transcription of cells due to their similarity with certain areas of recognition of primary transcription products. It seems that viroids prevent the correct cutting out of the introns. They are presumably multiplied with the aid of the cellular DNA-dependent RNA-polymerase II. Viroids occur mainly in warm climates and cause significant loss of profit as the causative agents of the potato disease or the Cadang-Cadang disease of palms.

The virus concentration within plant cells is high, although a virus like the TMV does not harm the host seriously. Infected cells contain often voluminous virus crystals. TMV was thus the first virus to be isolated in a pure state and in large amounts.

Plants are far from being defenceless against viruses. Only a few virus species are able to penetrate meristematic tissues or to infect a number of successive plant generations (vertical transmission). Hypersensitivity is an effective protection. It is based on the dying of cells in the immediate surrounding of the primary site of infection, thus stopping the spreading of the virus. In the case of Nicotiana tabacum, the genetic cause of hypersensitivity has been analyzed. One Nicotiana gene product guarantees hypersensitivity against all TMV strains, while another affects only some TMV strains. The symptom a virus causes at the primary site of infection is called primary symptoms. Symptoms caused by its spreading throughout the rest of the plant are called secondary symptoms.

Virus infections can usually be recognized by mosaic-like leaf patterns of light and dark green. The infection spreads often over the whole leaf beginning at the leaf veins. Leaves that had been infected during their development are usually deformed or involute. Frequently, lightened leaf areas, called chloroses, develop around the primary site of infection. Withered areas are called necroses. Chloroses are caused by a breakdown of the chlorophyll resulting in a decreased rate of photosynthesis. Heavy infections are characterized by a complete local loss of chlorophyll. Affected areas have a yellowish look as only the carotenoids remain. Some TMV strains, for example, can be recognized by yellow leaf areas ("yellow strains"). In nature, such strains are pretty rare since such a strong damaging of the host does also impair the virus' chances of replication and spreading.

Table 1.6. Plant viruses Carlavirus.

Name	*Accession Number*
Carnation latent virus	DSMZ PV-0365
Carnation latent virus	DSMZ PV-0366
Chrysanthemum virus B	DSMZ PV-0458
Cowpea mild mottle virus	DSMZ PV-0090
Garlic common latent virus	DSMZ PV-0449
Helenium virus S	DSMZ PV-0336
Kalanchoe latent virus	DSMZ PV-0290
Passiflora latent virus	DSMZ PV-0222
Poplar mosaic virus	DSMZ PV-0341
Potato virus M	DSMZ PV-0273
Shallot latent virus	DSMZ PV-0425
Shallot latent virus	DSMZ PV-0426

Some viruses multiply within the plant without causing symptoms. This phenomenon is called leten infection. In contrast, wound tumor virus causes the development of tumors. The symptoms of most viruses are dependent on both virus and host, and do thus present an important diagnostic feature.

G. Samuel proved in 1934 experimentally that the TMV virus uses the plant's vascular system for spreading, actually both the plant's transport to and way. As a result, fully differentiated, old leaves and roots, and young leaves are equally infected.

Since decades now, plant viruses (and especially the TMV) are favourite test objects of pure research due to the large amounts of material extractable from infected plants. The TMV was, too, the first biological electron-microscopic object to be photographed.

Table 1.7. Plant viruses Tymovirus.

Name	*Accession Number*
Belladonna mottle virus	DSMZ PV-0042
Eggplant mosaic virus	DSMZ PV-0060
Eggplant mosaic virus	DSMZ PV-0061
Eggplant mosaic virus	DSMZ PV-0062
Okra mosaic virus	DSMZ PV-0264
Poinsettia mosaic virus	DSMZ PV-0340
Turnip yellow mosaic virus	DSMZ PV-0299

Isolated TMV-RNA itself is infectious. It can thus be a carrier of genetic information. Chemical changes of the TMV-RNA (deamination of adenine or cytosine due to treatment with nitrite) showed in addition and for the first time that changes of the RNA are mutagenic, too. The changes of these first experiments caused mutations of a coat protein. The analysis of a large number of chemically induced mutants (performed by the working groups above yielded the first results important for the encoding of the genetic code. Work on the TMV made plain, too, that the gene-carrying RNA does not simultaneously function as mRNA. mRNA is synthesized within the cell as a separate fraction.

Table 1.8. Plant viruses Caulimovirus.

Name	*Accession Number*
Cauliflower mosaic virus	DSMZ PV-0226
Cauliflower mosaic virus	DSMZ PV-0227
Cauliflower mosaic virus	DSMZ PV-0228
Cauliflower mosaic virus	DSMZ PV-0229
Cauliflower mosaic virus	DSMZ PV-0362

Table 1.9. Plant viruses Potexvirus.

Name	*Accession Number*
''Tamus red mosaic virus''	DSMZ PV-0397
Cactus virus X	DSMZ PV-0083
Cymbidium mosaic virus	DSMZ PV-0334
Hydrangea ringspot virus	DSMZ PV-0223
Hydrangea ringspot virus	DSMZ PV-0246
Hydrangea ringspot virus	DSMZ PV-0372

Potato aucuba mosaic virus	DSMZ PV-0007
Potato virus X	DSMZ PV-0014
Potato virus X	DSMZ PV-0015
Potato virus X	DSMZ PV-0017
Potato virus X	DSMZ PV-0018
Potato virus X	DSMZ PV-0019
Potato virus X	DSMZ PV-0020
Viola mottle virus	DSMZ PV-0224
White clover mosaic virus	DSMZ PV-0310

Table 1.10. Plant viruses Cucumovirus.

Name	*Accession Number*
Cucumber mosaic virus	DSMZ PV-0035
Cucumber mosaic virus	DSMZ PV-0036
Cucumber mosaic virus	DSMZ PV-0037
Cucumber mosaic virus	DSMZ PV-0038
Cucumber mosaic virus	DSMZ PV-0183
Cucumber mosaic virus	DSMZ PV-0184
Cucumber mosaic virus	DSMZ PV-0185
Cucumber mosaic virus	DSMZ PV-0187
Cucumber mosaic virus	DSMZ PV-0188
Cucumber mosaic virus	DSMZ PV-0189
Cucumber mosaic virus	DSMZ PV-0314
Cucumber mosaic virus	DSMZ PV-0357
Cucumber mosaic virus	DSMZ PV-0358
Cucumber mosaic virus	DSMZ PV-0359
Cucumber mosaic virus	DSMZ PV-0418
Cucumber mosaic virus	DSMZ PV-0419
Cucumber mosaic virus	DSMZ PV-0434
Cucumber mosaic virus	DSMZ PV-0445
Cucumber mosaic virus	DSMZ PV-0453
Peanut stunt virus	DSMZ PV-0190
Peanut stunt virus	DSMZ PV-0203
Peanut stunt virus	DSMZ PV-0384
Tomato aspermy virus	DSMZ PV-0068
Tomato aspermy virus	DSMZ PV-0312

Table 1.11. Plant viruses Comovirus

Name	*Accession Number*
Andean potato mottle virus	DSMZ PV-0057
Andean potato mottle virus	DSMZ PV-0058
Andean potato mottle virus	DSMZ PV-0059
Broad bean true mosaic virus	DSMZ PV-0098
Cowpea mosaic virus	DSMZ PV-0074
Cowpea severe mosaic virus	DSMZ PV-0050
Radish mosaic virus	DSMZ PV-0306
Radish mosaic virus	DSMZ PV-0355

Table 1.12. Plant viruses Tombusvirus

Name	*Accession Number*
"Pelargonium virus, unnamed"	DSMZ PV-0371
Carnation Italian ringspot virus	DSMZ PV-0069
Cucumber necrosis virus	DSMZ PV-0313
Cymbidium ringspot virus	DSMZ PV-0272
Neckar river virus	DSMZ PV-0270
Pelargonium leaf curl virus	DSMZ PV-0265
Pelargonium leaf curl virus	DSMZ PV-0370
Petunia asteroid mosaic virus	DSMZ PV-0339
Tobacco necrosis virus + satellite	DSMZ PV-0198
Tobacco necrosis virus + satellite	DSMZ PV-0218
Tomato bushy stunt virus	DSMZ PV-0268
Tomato bushy stunt virus	DSMZ PV-0269
Tomato bushy stunt virus	DSMZ PV-0284
Tomato bushy stunt virus	DSMZ PV-0285

Double-Stranded RNA-Viruses; Wound Tumor Viruses

The genome of wound tumor viruses (WTV) consists of 12 double-stranded RNA segments. Neither of them is infectious. The wound tumor viruses of plants and animals (reoviruses) are related and are characterized by a number of common activities. They contain, for example, the enzyme transcriptase that transcribes single-stranded RNA or, in other words, produces an mRNA complementary to the transcribed strand. Each of the 12 segments can be transcribed and it is assumed that each of them encodes one protein. Replication occurs within the cytoplasm. Infection takes place via insects (e.g. aphids) that function both as a vector and an intermediate host, i.e. the virus multiplies in their tissue, too.

More than 50 plant species are known to be susceptible for wound tumor viruses. Among the symptoms are small tumors at the stem and larger and more numerous ones at the roots. WTV-induced wound tumors of leguminosae can be clearly distinguished from the nodules caused by nodule bacteria. In some plant species, like the lobelia, the infection induces the development of organs from otherwise normal organs, e.g. a leaf can develop at the lower leaf surface of another leaf.

Plant Viruses with Circular, Single-Stranded DNA: Gemini Viruses

The particles of gemini viruses are quasi-isometric. They are called gemini (twin) viruses, because they are usually found in pairs. Each particle has a diameter of just 15 - 20 nm. Gemini viruses belong to the smallest virus particles able to multiply without a helper virus. They have a circular DNA with a molecular weight of $0.7 - 0.8 \times 10^6$ (about 2,500 base pairs). In the case of some gemini viruses, it has been proven, in the case of others, it is assumed, that the genome consists of two molecules of DNA of almost equal size, but different sequence. The nucleotide sequences of some species are known.

Isolated circular DNA alone is not infectious. In infected host cells, the nucleus holds the chief amount of viral DNA. It is therefore assumed that the nucleus is also the place of its replication. It looks as if - analogous to the TMV - double-stranded intermediates would exist (= replication state: RF), since double-stranded DNA has been isolated from infected cells. Gemini viruses are of interest to genetic engineers, since they can infect monocots where their DNA enters the nucleus, too, and since the interest in monocot vectors is still large. The bean golden mosaic-virus (BGMV), the cassava latend-virus (CLV), the tomato golden-mosaic-virus (TGMV), the maize-streak virus (MSV), and the abutilon mosaic-virus belong all to the Gemini-virus family.

Insects (greenhouse whitefly, grasshoppers, and others) help usually in spreading Gemini-viruses in nature. These viruses can cause considerable damage to agriculture.

Double-Stranded DNA Viruses

The prototype of a plant virus with double-stranded DNA is the cauliflower mosaic-virus (CaMV). Double-stranded plant viruses, too, have been considered in genetic engineering. It is tried to convert them into vectors in order to put foreign DNA into plant cells similar to the way the plasmid of Agrobacterium tumefaciens is used.

CaMV-DNA is normally not incorporated constantly into the host cell genome (R. Sphepherd and R.J. Wakeman, 1971). CaMV is the collective name for a group of tightly related viral species usually transmitted by aphids. Each species has a narrow host specificity, overlapping with that of other species is rare. The virion is spherical with a diameter of about 50 nm. Most likely, the capsid consists of 420 identical subunits with molecular weights of 42,000. The DNA contains about 8,000 base pairs. In the case of one species, it has been sequenced and 8,024 base pairs were found.

Neutron diffraction experiments showed that the DNA is sandwiched between the protein subunits. The core of the viral particle contains no proteins or nucleic acids. One of the three ring-shaped DNA molecules is broken up in three different places. Treatment of the viral DNA with SI-endonuclease, a DNAse breaking down single-stranded DNA, renders therefore three fragments of defined length. Further analyses showed that the ring is no closed, uniform structure, but that it results from the combining of three molecules kept together at their ends by complementary sequences.

Symptoms occur two to three weeks after infection and can be recognized by the mosaic-like lesions of infected leaves. The virus spreads systemically, its secondary symptoms are similar to those of the primary infection. Leaves that were infected during their development display deformed leaf blades.

Viroids, the Smallest Infectious Units

Viroids are infectious units that cause a number of plant diseases. They are circular molecules of RNA with molecular weights between 107,000 and 127,000. H. J. Gross et al. sequenced the nucleotide sequence of the potato spindle tuber virus (PSTV) in 1978. It consists of 359 ribonucleotides and is characterized by numerous intramolecular base-pairings that lend stability to the structure. They are organized in a sequence of helices separated from each other by loops. The resulting structure resembles a dumb-bell with an axis ratio of 1:20. Several more viruses have been sequenced in the meantime. All of them have structures similar to that of the PSTV.

They are ~240 - 380 nucleotides long and all of them have dumb-bell structures. The fact that a central portion of the molecule that is responsible for the pathogenicity of the viroids, is structurally conserved is especially interesting. PSTV-mutants with changes in this section are less pathogen than the wild type. Viroids multiply even at relatively high temperature (about 35°C). Most likely, they have adapted to their host plants that have so-far strictly been found to inhabit tropical, subtropical, and continental climates. The viroids are localized within the chromatin fraction of the nucleus. The DNA-dependent RNA-polymerase II and I use the viroids as templates and produce strands that again serve as templates for the synthesis of the +strand.

2

VIRAL DISEASES

People have been combating weeds, insects, and plant diseases throughout history. Pests of the plants man use for food are abundant. Pests (insects and weeds), and diseases, developed "hand in hand" with the plants they attacked. Pests are plants, animals or viruses that are detrimental to humans and crops. Cultivation of one or a few crops species on the same land for several years or decades shifts the population balance of soil microorganisms, plant pathogens, insect pests, and weeds to those strains favored by the crops grown. Most crops are subject to attack from a large number of pests (insects, weeds) and diseases. These are considered to be the major limiting factor in crop production. The susceptibility of cultivars to attack from specific pests or diseases varies greatly. Development of resistant lines would seem to offer the best prospects of crop protection, particularly if this is combined with the use of healthy, clean seed, efficient weed control and crop rotation. It is very difficult to make reliable estimates of losses due to pests. There is no question, however, that these losses are of substantial and often staggering proportions. A conservative estimate suggested that plant pests destroyed one third of mankind's supply of food and fiber every year. Losses of much greater magnitude occur as a matter of course in many less developed countries. It is only in recent times that man's response to pest control has been based on any appreciable understanding of the nature and causes of pest problems. Methods of pest control change and become increasingly effective as we gain greater understanding of pests and their habits. Although there remains much that we do not know, we can formulate pest control programs on a rational basis. A control program should be based on an understanding of the biology and habits of the pest, a consideration of all effective methods of control, and recognition of the level of control that is both desirable and possible. Some of the preventive measures and control methods for pests and diseases are discussed in this lesson.

Plants and plant products are tremendously important for human survival since they provide food, clothing, furniture, a stable environment, and often housing. Plants, whether cultivated or wild, generally grow well when the soil provides them with sufficient nutrients and moisture, sufficient light reaches their leaves, and the temperature stays within a "normal" range. However, like people, plants can get sick. Agents similar to those that cause disease in people also can cause diseases in plants. The broadest definition of plant disease includes anything that damages plant health. This definition can include such diverse factors as pathogens, insufficient nitrogen, air pollution, lawnmower damage, and deer damage. A stricter definition usually includes any persistent irritation resulting in plant damage and characteristic symptoms. This definition includes such factors as pathogens, insufficient nitrogen, and air pollution.. However, it excludes factors such as lawnmower injury to trees and lightning injury since this damage presumably is a one-time occurrence. A very strict definition includes only infectious organisms (pathogens) that multiply and spread to other nearby plants. Most pathogens are microscopic and include bacteria, fungi, nematodes, viruses, mollicutes, protozoa, and parasitic plants. These are called biotic diseases. Plant pathologists usually do not consider organisms such as insects, deer, rodents, and birds to be pathogens.

As a science, plant pathology strives to increase our knowledge about plant diseases. To do this, plant pathologists study such factors as the biology of the pathogenic organisms, the mechanisms that pathogens use to cause disease and interactions between pathogens and host plants. These studies help plant pathologists to devise better methods of preventing or controlling diseases and alleviating the damage that they cause.

In order for a biotic plant disease to occur, three conditions must be met:

1. The host plant must be susceptible.
2. A pathogen must be present.
3. The environment must be favorable for infection by the pathogen.

All three of these factors must occur simultaneously. If one or more is absent, the disease does not occur. The genetics of the plant determine its susceptibility or resistance to a particular disease. Specific pathogens vary in their ability to infect different plant species. Various physical and biochemical factors of a given plant species influence susceptibility. These factors include plant defense mechanisms, carbohydrates and protein types, cuticle thickness, and stomatal shape. The developmental stage of the plant also can influence disease development. Pathogens differ in their ability to survive, spread, and reproduce. For instance, a certain viral disease may occur only when a specific insect vector transports virus particles to the susceptible plant. Certain temperature, light, or moisture conditions generally are necessary for a disease to occur.

Since the first time modern plant breeding began, in-corporation of genes conferring pathogen resistance into commercial varieties has been a goal of importance. To get satisfactory result requires that in the plant genomes, resistance genes exist. It also requires that the resistance genes be compatible with the crop variety being developed. Once introduced by a series of crosses the resistance genes must be effective and have good resistance to the target problem. For example, the N gene, which was resistant to tobacco, fifty years ago from a tobacco relative, Nicotiana glutinosa and this virus-resistance is still widely used effectively. Most of the time however, resistance doesn't last. The population of the viral stains, where there is a race between plant breeders using new resistance genes and the pathogens that are targeted, overcomes resistance. Many crop plants are affected by viral disease, for which there is no source of resistance in compatible plant genotypes. A lot of people are unaware of the impact of plant viruses. All over the world serious problems from plant viral diseases come up. That's why, when the first virus-resistant transgenic plants were described, the workers in the field were definitely excited. A major ecological risk associated with the widespread cultivation of transgenic crops is genetic flow of the transgene to non-transgenic populations of the same, or other, species. If genetic flow to viral population could occur in nature, and fit recombinant viruses could spread in the field. Weather this possibility is more probable than or as probable as that of similar recombinant viruses originating from mixed infections in non-transgenic plants is a point that needs to be evaluated. Recombinant viruses can sometimes be hard to fight. Geminiviruses transmitted by the white fly (begomoviruses) is a good example of emerging viruses in the plant kingdom. Virus groups have analyzed the emergence and in some instances recombination between the pre-existing white flies and new ones seem to be at the root of the appearance of new viruses and stains that cause devastating epidemics. Infecting viruses are a big concern since they could lead to recombinant viruses with modified biological properties. RNA recombination occurs naturally in viral genomes, which there is clear evidence to prove it. It is generally considered that, mutation, assortment, and recombination are three main molecular forces that increase variability of the revolution of RNA viruses. Plant viral disease dangers can be devastating to farmers, ranchers, and the surrounding communities.

Viruses are so small that they cannot be seen with an ordinary microscope. Their effects on plants generally recognize them. Often it is difficult to distinguish between diseases caused by viruses or mycoplasmas and those caused by other plant disease agents such as fungi and bacteria. Usually, the best way to identify a virus is to compare the symptoms with pictures and descriptions of diseased plants for which a positive identification has been made. Other methods require more sophisticated testing, such as inoculating indicator plants and observing the results or using specifically identified antibodies to test for the presence of the organism.

Viruses depend on other living organisms for food and to reproduce. They cannot exist separately from the host for very long. Viruses are commonly spread from plant to plant by mites and by aphids, leafhoppers, whiteflies, and other plant- feeding insects. They may be carried along with nematodes, fungus spores, and pollen, and may be spread by people through cultivation practices, such as pruning and grafting. A few are spread in the seeds of the infected plant.

Viruses can induce a wide variety of responses in host plants. Most often, they stunt plant growth and/or alter the plant's natural colour. Viruses can cause abnormal formation of many parts of an infected plant, including the roots, stems, leaves, and fruit. Viruses usually cause Mosaic diseases, with their characteristic light and dark blotchy patterning. Viroids are similar to viruses in many ways, but they are even smaller and lack the outer layer of protein that viruses have. Only a few plant diseases are known to be viroid-caused, but viroids are the suspected cause of many other plant and animal disorders. Viroids are spread mostly through infected plant stock. People can spread infected plant sap during plant propagation and other cultural practices. A few viroids are known to be transmitted with pollen and seeds. Mycoplasmas are the smallest known independently living organisms. They can reproduce and exist apart from other living organisms. They obtain their food from plants. Yellows diseases and some stunts are caused by mycoplasmas. Insects spread most mycoplasmas, most commonly by leafhoppers. Mites may also spread them. Mycoplasmas are also readily spread among woody plants by grafting.

As discussed in previous chapters, virus is an extremely small organism composed of genetic material (either RNA or DNA) surrounded by a protein coat. Viruses replicate only inside living cells. Viruses are able to take command of plant cell metabolism and direct the cell to manufacture more virus particles. An average virus is 100,000 to 1,000,000 times smaller than a plant cell. Due to their size and biology, plant viruses traditionally have been difficult to identify. Symptoms give important clues, but yield no real proof of the exact pathogen involved. In recent years, many advances in molecular biology techniques have allowed plant pathologists to "label" viruses in order to identify them. Virologists have developed many different labeling techniques for virus identification. There is some example of one such technique. For detecting tobacco mosaic virus (TMV) in tomato, virologists have developed a substance that turns blue only when a solution of crushed plant material containing TMV is present. If the tomato spotted wilt virus (TSWV) is present instead, the solution remains clear. To confirm the presence of TSWV, one must conduct a second test containing a substance that turns color in the presence of TSWV. It would be too time consuming to run tests for the 30 possible tomato viral diseases. This is where the art of plant disease diagnosis is important. A skilled diagnostician can look at the plant symptoms, make guesses as to the viruses present, and then run tests specific for these viruses to scientifically identify the disease. Through previous chapters, it is cleared that viroid is a piece of genetic material that, unlike a virus, has no protein coat. Viroids divert plant metabolism to produce more viroids. They spread by vegetative propagation.

Current Strategies Used to Control Plant Viruses

1. The principle method is to identify resistance genes in a species and incorporate these genes into cultivars and varieties that have other good agronomic or horticultural properties. In some cases there are single genes that provide resistance to a particular virus. In others, there may be a small number of genes that in combination provide resistance. These are known as host resistance genes. In the last two years, the first of these genes for resistance to a virus, in this case against tobacco mosaic virus, has been cloned.

2. A second method to control viruses is by controlling how they are spread. For viruses that are spread mechanically, that means minimizing contact between healthy and infected plants, and anything that is shared between them, such as greenhouse pruning tools. Eliminating smoking in greenhouses is apparently important to controlling the spread of TMV, which is present in cigarette tobacco.
3. For viruses transmitted by vectors, controlling the population of the vector is another method of control. This can include cultural practices. For example, it is recommended that growers delay planting winter wheat until after the "fly free day". Not only does this reduce the damage that can be caused by the Hessian fly, it also prevents infection of the wheat seedlings with Barley Yellow Dwarf Virus (BYDV), which is transmitted by aphids. Aphids are no longer active after the "fly free day" and so cannot infect the crop in the fall. The wheat can still be infected by BYDV in the spring, but a later infection is less likely to cause serious losses.
4. Another method to control virus infections in some crops is called cross-protection. This is used in tomatoes and citrus crops. Plants are deliberately infected with a mild strain of a virus, which produces only mild symptoms. The plants that have been infected with the mild strain of the virus are in some way protected against a subsequent infection by a more severe strain of the virus. Plants do not have an immune system similar to that of mammals and this is not the same as a vaccination that animals might have received but the outcome is similar. The mild strain of virus protects the plant from a subsequent infection. This formed the basis for experiments to examine this phenomenon and led to the development of a general method to make plants resistant to viruses. The hypothesis was put forward that some component of the virus was responsible for triggering this resistance phenomenon.

The Life Cycle of a Typical Plant Virus

1. When a virus enters a cell, the RNA is released from the protein coat.
2. This RNA is then used as template to make RNA and proteins encoded by the virus. The virus uses the machinery of the cell to express the genetic information carried by the viral genome.
3. The genome of the virus is replicated and new coat proteins are synthesized. This leads to the assembly of new virus particles in the infected cells.

4. Finally the virus is transmitted to another plant, and this can be done in a number of ways:
   (a) Mechanically by physical touching of infected and healthy plants; for example, in greenhouses some viruses are spread by using the same tools, such as pruners, which carry virus from infected plants to those that are healthy. Good sanitary practices can reduce the spread of viruses by pruning tools.
   (b) By a biological vector, such as aphids or other insects that feed on an infected plant, ingest some virus particles, then move to a healthy plant and transmit the virus in the saliva when feeding.
   (c) Some viruses are carried in the seed and so are passed from one generation to another

Symptoms and Viral Pathogenesis

Type of Viral Symptoms

Symptoms of viral diseases were well described even before viruses were well characterized. In early days when techniques for virus purification and visualization were not available, virus detection and identification were entirely based on symptoms. For this historic reason, most virus names were derived from the symptoms they cause on host plants. In most cases, you can tell typical symptoms of a virus from its name. For example, tobacco mosaic virus (TMV) obviously cause mosaic symptom on tobacco plants.

Virus infection often alters the host metabolism and interrupts important cellular functions. The effect of these alterations is expressed macroscopically as deviations from normal morphology. When a plant does not support virus replication or translocation, no viral symptoms are induced. These plants are called resistant plants. Susceptible plants are the plants that allow infection and movement of viruses.

A combination of several symptom types usually shows up in a plant infected by a virus. For example, wheat or barley plants infected by barley yellow dwarf virus are dwarfed and all the leaves turn yellow in color. For the following description, we will consider each symptom type individually.

No obvious symptoms

Not all virus infections would necessarily cause visible symptoms on infected plants. A virus can multiply to a very high concentration and yet induce no obvious symptoms in an infected plant. Infection with a mild strain of the virus in a tolerant host will result in a symptomless infection. A classical example is TMV on Ambalema tobacco. TMV infects and replicate throughout tobacco plants, but infected plants show no signs of disease. A more practical example is citrus tristeza virus (CTV). Citrus tristeza virus causes quick death of citrus on sour orange rootstock. In Meyer lemon tree, CTV replicates to a high titer and the tree remain normal in appearance. Plants that are infected with viruses but do not display symptoms are called symptomless carriers. Symptomless carriers play a significant role in a virus epidemic. Two genera from the family of Partiviridae: alphacryptovirus and betacryptovirus infect plants but do not induce obvious symptoms on infected plants. This group of viruses have double-stranded RNA genome. They are transmitted efficiently though pollen or seeds but are not transmissible mechanically or by vectors. These viruses may have co-evolved with their hosts and become almost a part of the host genetic materials.

Localized symptoms (local lesions)

Local lesions usually develop near the site of entry on the leaves or fruit. Most types of symptoms see are chlorotic or necrotic spots, ringspots, ranging from single rings from concentric rings of rings of chlorotic or necrotic nature. Local symptoms are usually a result of interaction between host defense responses and virus infections. These defense responses include rapid death of cells (hypersensitivity) and other physiological changes, which limit virus spread further. Viruses are limited to local lesions most of the times. However, they may spread out and cause systemic infections even if there is a rapid death of cells near the inoculation site. Ringspots with several concentric rings are probably the best documentary of the struggle between plants and viruses.

1. *Chlorotic lesions*: Lesions are discolored or pale. Cells in the lesion are apparently alive. How viruses are restricted in this type of lesions is not very clear.
2. *Necrotic lesions*: Lesions consist of dead cells. This is the result of typical hypersensitive reaction. However, viruses do occasionally escape the local lesion and lead to systemic infection.
3. *Ringspot*: Ringspot is a ring-like lesion. Spectacular ringspots with several concentric rings are often observed.

Systemic symptoms

Systemic symptoms refer to symptoms that develop throughout the plants. They are the results of systemic invasion of the plant host by a virus. When viruses reach the vascular tissues, viruses are transported throughout the plants and causes systemic infection. Depending on the appearance of the symptoms they can be classified as the following:

1. *Mosaic patterns*: Mosaic patterns are more common symptoms in virus infection. Mosaic simply means an alternating pattern of tissues with two different colors. The colors can be light green, dark green, yellow, golden, or even white. If the borders between areas are sharp, then the symptom is referred as mosaic, if the borders are defused, it is called mottle. There are no definite distinctions between the two. Other types of related symptoms are vein clearing or vein banding. The mosaic type symptoms can occurs on leaves, fruit or flowers. On monocotyledon plants, the symptoms are more likely to be of stripes or streaks. The extreme mosaic type of symptoms is yellowing of most of the infected leaves, as in bean golden mosaic virus.

   Dark green islands (DGIs) are a common symptom of plants systemically infected with a mosaic virus. DGIs are clusters of green leaf cells that are free of virus but surrounded by yellow, virus-infected tissue. First, transcripts of a transgene derived from the coat protein of Tamarillo mosaic potyvirus (TaMV) were reduced in DGIs relative to adjacent yellow tissues when the plants were infected with TaMV. Second, nontransgenic plants coinfected with TaMV and a heterologous virus vector carrying TaMV sequences showed reduced titers of the vector in DGIs compared with surrounding tissues. DGIs also were compared with recovered tissue at the top of transgenic plants because recovery has been shown previously to involve PTGS. Cytological analysis of the cells at the junction between recovered and infected tissue was undertaken. The interface between recovered and infected cells had very similar features to those surrounding DGIs. They conclude that DGIs and recovery are related phenomena, differing in their ability to amplify or transport the silencing signal.

2. *Yellowing*: Infection by some viruses may induce yellowing throughout the infected plants. Viruses that cause this type of symptoms are usually phloem limited. Infection of the phloem tissue may disrupt transportation of nutrient and water throughout vascular tissue, resulting general yellowing.

3. *Dwarfing*: Stunting and dwarfing of the infected plants are another common symptom of virus infections. The degree of dwarfing varies. As a general rule, the size of the infected plants is almost always

smaller than a healthy one. Interference with plant growth hormone levels is regarded as the most important factor for dwarfing. Other factors include reduced photosynthesis and reduced translocation of nutrients.

4. *Necrosis*: Occasionally, an entire plant infected by a virus may become necrotic. Symptoms usually starts along the veins and whole leaves or plants may die eventually. This type of symptoms is usually caused by a delayed hypersensitivity reaction. The infecting virus has moved out of the entry sites and systemically invaded the entire leaf or plant before the hypersensitivity reaction kicks in.
5. *Malformation*: Change the normal shape of a plant or parts of a plant. Symptoms may include stunting for the whole plant, leaf malformation, twisting, curling, shoestring like leaves, enations on leaves, Gall or tumors formation, rolled either upward or downward.

Interference in Host Metabolisms

Viruses are intercellular parasites with a limited genome, and therefore dependent on the host cell for multiplication and proliferation. Viruses exert negative effects on the hosts by interfering with the host metabolism. Symptoms caused by viral infections are generally considered as results of the following: host defense response to viral infection (chlorotic or necrotic lesions), interference with hormone metabolisms (malformation and dwarfing) and nutrient transport (yellowing and dwarfing), reduced photosynthesis (mosaic, yellowing, and dwarfing), or a combination of the above. However, multiplication of viruses within host cells alone appears to not directly cause diseases. Some viruses may multiply to a high level in the host plants, and yet cause no obvious symptoms..

Hypersensitive reactions

Plants have developed several mechanisms to defend themselves against pathogens. One type of reaction involves rapid cell death around the entry site of a pathogen. This hypersensitive reaction (HR) results in visual symptoms as necrotic lesions in tissues localized around the site of the entry or in entire plants if the hypersensitive reaction is delayed until viruses have spread throughout the plant.

Specific recognition between viral genes and host genes is usually involved in this type of reaction. TMV coat protein gene and the N' gene controlled HR in tobacco is a good example. Severe strains of

TMV can systemically infect tobacco plants with N' gene while mild strains induce local lesion (HR) on plants with N' gene. Amino acid substitutions in the coat protein gene were observed in chemically induced mutants of the severe strains of TMV that induce HR phenotypes and in the mutants of mild strains that systemically infect N. sylvestris tobacco. Introduction of a specific point mutation in to TMV U1 strain (HR on N' plants) cause systemic infection of the engineered mutants on plants with N' gene.

Effect on chloroplasts and photosynthesis

Chloroplasts are important factories in green plants. They generate, through photosynthesis, energy and carbohydrate materials essential for plant growth and development. Some viruses, such as turnip yellow mosaic virus, induce obvious clumping and fragmentation of the chloroplasts, and formation of large vesicles. Viral capsid proteins of TMV and cucumber mosaic virus have been reported to enter into chloroplasts. Altered chloroplasts may be responsible for chlorosis or chlorotic type of symptoms and retarded growth of the infected plants. However, chlorosis induced by cucumber mosaic virus is not accompanied by obvious chloroplast damage. The alteration of cell organelles in wheat leaf cells systemically infected with wheat streak mosaic virus was quantitatively investigated at both light- and electron-microscopy levels. Nuclei significantly increased in size at an early stage of infection and contained dispersed heterochromatin. The nuclear envelope was frequently invaginated and had formed membrane-bound vesicles containing membranes, ribosomes, cylindrical inclusions, virus particles, and, occasionally, fibrils and mitochondria. Chloroplasts in infected tissue were smaller, frequently had extrusions, and contained far less starch than chloroplasts in healthy tissue. Many chloroplasts formed double membrane-bound invaginations in the envelope membrane. Virus particles and mitochondria were found within the invaginations. The viral infection also caused the chloroplast envelope's inner membrane to proliferate and fold repeatedly, dividing the stroma into numerous single membrane-bound fragments. Double membrane-bound elongated tubular structures accumulated in large quantities within the cytoplasm of infected cells.

Disruption of hormone balance

Several hormones are very important in regulating plant development such as cell divisions and cell elongation. Disturbance in the hormone balance results in malformation in plant parts, such as tumor formation, stunted growth, and leaf curling.

Nutrient uptake and translocation

Several phloem-limited viruses induce general yellowing of the entire plants, yet these viruses are not distributed throughout the leaf mesophyll and parenchyma cells. Viral interference of the phloem function such as nutrient uptakes and movement of nutrient from one part of plants to another probably affect this symptom type.

Interaction of Viral Proteins with Translational Apparatus

Several viral proteins are capable of interacting with proteins from the translation apparatus such as initiation factors eIF4E (Leonard et al., 2000; Schaad et al., 2000) and poly A binding protein (Wang et al., 2000). These interactions have the potential of interfering with host protein translation system, thus altering the gene expression pattern. Leonard et al. (2000) reported the interaction between the viral protein linked to the genome (VPg) of turnip mosaic potyvirus (TuMV) and the translation eukaryotic initiation factor eIF(iso)4E of Arabidopsis thaliana and Triticum aestivum (wheat). eIF(iso)4E binds the cap structure m7GpppN of mRNAs and has an important role in the regulation in the initiation of translation. The interaction domain on VPg was mapped to a stretch of 35 amino acids, and substitution of an aspartic acid residue found within this region completely abolished the interaction.

The cap analogue m(7)GTP, but not GTP, inhibited VPg-eIF(iso)4E complex formation, suggesting that VPg and cellular mRNAs compete for eIF(iso)4E binding. The biological significance of this interaction was investigated. Brassica perviridis plants were infected with a TuMV infectious cDNA (p35Tunos) and p35TuD77N, a mutant which contained the aspartic acid substitution in the VPg domain that abolished the interaction with eIF(iso) IE. After 20 days, plants bombarded with p35Tunos showed viral symptoms, while plants bombarded with p35TuD77N remained symptomless. These results suggest that VPg-eIF(iso)4E interaction is a critical element for virus production.

Similar interaction between NIa protein and eIF4E was also reported. NIa of potyviruses has been shown to confer host genotype-specific movement functions in plants. Specifically, NIa from tobacco etch virus (TEV)-Oxnard, but not from most other strains, confers the ability to move long distances in Nicotiana tabacum cultivar "V-20." To identify cellular proteins that interact with Nla in a host- or strain-specific manner, Schaad et al. (2000) searched a tomato cDNA library using a yeast two-hybrid system. Ten proteins that interacted with NIa were recovered, with translation initiation factor eIF4E being by far the most common protein identified. Interaction of elF4E with NIa was shown to be TEV strain-specific. eIF4E from both tomato and tobacco interacted well with NIa from the HAT strain, but not from the Oxnard strain.

Using similar yeast two hybrid searching, Wang et al. (2000) identified host plant proteins associating with the RdRp of zucchini yellow mosaic potyvirus (ZYMV). Several cDNA clones representing a single copy of a poly-(A) binding protein (PABP) gene were isolated from a cucumber (Cucumis sativus L.) leaf cDNA library. Deletion analysis indicated that the C-terminus of the PABP is necessary and sufficient for interaction with the RdRp. This is the first report of an interaction between a viral replicase and PABP and may implicate a role for host PABP in the potyviral infection process.

Change of Host Gene Expression through Gene Silencing and DNA Methylation

RNA silencing (post-transcriptional gene silencing) and DNA methylation have been suggested as host defenses against invasion of foreign nucleic acids including viruses. While these mechanisms work well in silencing the invading nucleic acids, they might also silence host genes with homologies to the invading nucleic acids. Both RNA-directed DNA methylation (Wassenegger et al., 1994; Pelissier and Wassenegger, 2000) and DNA-directed DNA methylation have been reported. The following example describes an RNA-directed DNA methylation. Wassenegger et al (1994) introduced one monomeric and three oligomeric potato spindle tuber viroid (PSTVd) cDNA units into the tobacco genome.

Southern analysis of the transgenic plants revealed that only their PSTVd-specific sequences become fully methylated. Viroid cDNA methylation could only be observed after autonomous viroid RNA-RNA replication had taken place in these plants. These findings demonstrate that a mechanism of de novo methylation of genes might exist that can be induced and targeted in a sequence-specific manner by their own mRNA. A recent study by Pelissier and Wassenegger (2000) showed that DNA with as little as 30-nucleotide sequence complementarity to PSTVd can be methylated upon infection by the viroid. Although this sequence specific methylation is only limited to the viroid DNA in the genome in this study, it suggests another mechanism that viroids can cause diseases. Production of viral or viroid RNA (especially at a high concentration) may turn off host gene(s) with localized homologous sequences.

Inhibition and Stimulation of Host Gene Expressions

Viral infections have been shown to both inhibit and stimulate host gene expressions. The best studied system in this respect is the infection of pea seed-borne mosaic virus (PSbMV) in pea cotyledon tissues. Wang and Maule (1995) reported that replication of PSbMV in the tissues could transiently inhibit host gene expression. Replication of PSbMV was restricted largely to a zone of cell close to the infection front. In that replication zone, expression of normal host mRNA was shut down completely. However, host mRNAs were expressed in virus-infected tissues outside of the replication zone. By examining the front of virus invasion in immature pea embryos, two heat-inducible genes encoding HSP70 and polyubiquitin were induced coordinately with the onset of virus replication. At the same time, two genes encoding lipoxygenase and heat shock cognate protein were down regulated. The down-regulation was part of a general suppression of host gene expression that may be achieved through the degradation of host transcripts. Their results indicate that, in contrast to some animal virus infections, there is not a general inhibition of translation of host mRNAs following PSbMV infection. This selective control of host gene expression was observed in all cell types of the embryo and identifies mechanisms of cellular disruption that could act as triggers for symptom expression. Similar host gene induction and down regulation were also observed with Pea early browning tobravirus, White clover mosaic potexvirus, and Beet curly top geminivirus.

Interference with Cell Cycle

The geminivirus contains single-stranded DNA genome and replicates in differentiated plant cells. These cells have left the cell division cycle and contain no detectable levels of DNA replication enzymes or proliferating cell nuclear antigen (PCNA) which associates with DNA polymerase delta and promote its processivity. Nagar et al. (1995) showed that tomato golden mosaic virus (TGMV) induces the accumulation of PCNA in mature cells of Nicotiana benthamiana. Later on, Egelkrout et al. (2001) determined that PCNA protein accumulation reflected transcriptional activation of the host gene. The PCNA transcript of approximately 1200 nucleotides was detected in young leaves, but not in mature leaves of healthy plants. However, the same transcript was found in mature leaves of infected plants. Analysis of the PCNA promoter showed that it was induced during TGMV infection. Developmental studies established a strong relationship between symptom severity, viral DNA accumulation, PCNA promoter activity, and endogenous PCNA mRNA levels. Together, their results demonstrate that geminivirus infection induces the accumulation of a host replication factor by activating transcription of its gene in mature tissues, most likely by overcoming host-mediated repression.

Meanwhile Xie et al. (1995) revealed how wheat dwarf geminivirus AL1 protein might ractivate the host DNA replication machinery by interacting with the retinoblastoma protein (pRb), a cell cycle control protein. Unlike other DNA virus proteins, AL1 does not contain the pRb binding consensus, LXCXE, and interacts with plant pRb homo logues (pRBR) through a novel amino acid sequence. Kong et al. (2000) mapped the pRBR binding domain of AL1 between amino acids 101 and 180 and identified two mutants that are differentially impacted for AL1-pRBR interactions. Plants infected with the E-N140 mutant, which is wild-type for pRBR binding, developed wild-type symptoms and accumulated viral DNA and AL1 protein in epidermal, mesophyll and vascular cells of mature leaves. Plants inoculated with the KEE146 mutant, which retains 16% pRBR binding activity, only developed chlorosis along the veins, and viral DNA, AL1 protein and the host DNA synthesis factor, proliferating cell nuclear antigen, were localized to vascular tissue. Their results established the importance of AL1-pRBR interactions during geminivirus infection of plants.

Symptoms Induced by Transgenic Expression of Single Viral Genes

With the technological improvement in plant transformation, more and more viral genes are being inserted into plant genomes for one reason or another. Most of viral genes did not confer any phenotypical changes in the transgenic plants. However, some did results in symptoms similar to those of viral infections. Transgenic Nicotiana benthamiana plants expressing squash leaf curl virus (SqLCV) BL1 movement protein gene were abnormal both in their growth properties and phenotypic appearance (Pascal et al., 1993). Leaves from the transgenic plants were mosaic and curled under, symptoms typical of SqLCV infection in this host. On the other hand, transgenic plants expressing movement protein BR1 or a truncated BL1 were phenotypically indistinguishable from wild-type N. benthamiana. BL1 was localized to the cell wall and plasma membrane fractions, whereas BR1 was predominantly in the microsomal membrane fraction. These findings demonstrate that expression of BL1 in transgenic plants is sufficient to produce viral disease symptoms. This observation was later confirmed with tomato mottling geminivirus (TMoV). Tobacco plants transformed with the TMoV movement protein (pathogenicity) gene (BL1) had phenotypes ranging from plants with severe stunting and leaf mottling (resembling geminivirus symptoms) to plants with no visible symptoms. The sequence data for the BC1 transgene from the transgenic plants with the different phenotypes indicated an association of spontaneously mutated forms of the BC1 gene in the transformed tobacco with symptomatic phenotype variations. Both of the above examples involve viral movement proteins, which are known to interact and increase the size exclusion limit of plasmodesmata. Perhaps the increased size exclusion limit interferes with the movement of signal molecules or even large RNA and protein molecules between plants cells, resulting altered metabolisms. Virus-like symptoms were also reported in transgenic plants expressing the open reading frame 0 (ORF0) of luteoviruses. Van der Wilk et al. (1998) constructed transgenic plants expressing the potato leafroll luteovirus (PLRV) ORF0 p28 protein. ORF0 expressing transgenic potato plants displayed an altered phenotype resembling virus-infected plants. A positive correlation was observed between levels of accumulation of the transgenic transcripts and severity of the phenotypic aberrations observed. In contrast, potato plants transformed with a modified, untranslatable ORF0 sequence were phenotypically indistinguishable from wild-type control plants. Southern blot analysis showed that the transgenic plants that accumulated low levels of ORF0 transcripts detectable only by reverse transcription-polymerase chain reaction, contained methylated ORF0 DNA sequences, indicating down-regulation of the transgene provoked by the putatively unfavorable effects p28 causes in the plant cell.

Viral Recombination in Virus Resistant Transgenic (VRT) Plants

Plant viruses contain only a few genes: a replicase, coat protein, movement protein, perhaps a few others. The genome of the virus is comprised of RNA. In order to construct chimeric genes that direct the expression of individual viral proteins in transgenic plants, the RNA first had to be converted into a DNA copy. An RNA sequence can be copied into a DNA molecule, maintaining the correct sequence of bases, by reverse transcriptase. This is the first step towards making the chimeric genes to express viral proteins, as outlined below.

1. Copy the viral RNA into DNA using reverse transcriptase, make into a double stranded DNA.
2. Clone this DNA into a pslamid and sequence the viral genome to identify the genes in the virus.

3. Construct a chimeric gene consisting of the open reading frame for the coat protein and a strong promoter, such as the 35S promoter, to give high-level expression of this transgene.
4. Produce transgenic plants containing this transgene.

This was first tested with transgenic tobacco plants engineered to express the coat protein of tobacco mosaic virus (TMV). When transgenic tobacco plants that express the coat protein of TMV were exposed to the virus, they were found to be resistant to infection by the virus. The inoculated transgenic plants did not develop symptoms, while the control non-transformed plants did become infected. However, the transgenic plants expressing the TMV coat protein were not completely immune; they would become infected if inoculated with a high dose of the virus, but it took longer for symptoms to develop and the symptoms were less severe.

This protection works against the virus from which the coat protein was isolated, as well as against closely related viruses. Having been shown to work providing protection against TMV in tobacco, the same strategy (expression of the virus coat protein in transgenic plants) was then applied to other plant viruses. This approach has now been shown to work providing resistance against a large number of different plant viruses. This is called coat protein-mediated resistance.

In this is not exactly clear, but it appears to block some early stage in the infection process, maybe the uncoating of the virus when it first enters the plant cell. If transgenic plants are infected expressing coat protein with normal viruses, the plants are protected, but if the plants are infected with the RNA of the virus, the plants will develop symptoms. So it appears that expressing the coat protein in the plant prevents the virus from uncoating, expressing the viral genome and causing disease symptoms.

This method has been used successfully in greenhouse and field tests to obtain resistance to many plant viruses. While the first, and the majority, of experiments and trials have been conducted with model systems such as tobacco, the strategy appears to be equally successful with crop plants. It appears to be a technique, which can be applied very widely to obtain resistance. Another vegetable crop that has been developed with resistance to a virus using coat protein mediated resistance is potato. These are being developed by Monsanto with resistance to potato leafroll virus and potato virus Y. Perhaps the best example to date of this strategy is in trying to develop papaya trees with resistance to papaya ringspot virus (PRSV). This is essentially a lethal diseasefor papaya and prevents commercial production of these tropical fruits. There was a thriving papaya industry in Hawaii, which had to move from one region of the island to another because of the spread of PRSV.

Over the last ten years researchers at a number of institutions have been developing transgenic papaya that express the coat protein of PRSV. Clone a cDNA for the PRSV coat protein gene. Construct a chimeric gene where the open reading frame of this coat protein is placed under the control of a high level promoter. Transfer this chimeric gene into papaya by transformation. Regenerate transgenic plants and evaluate for resistance. This has led to the release of a transgewic papaya with resistance to PRSV. which is now being planted in Hawaii. While coat protein-mediated resistance has proven to be quite effective, other methods have also been tested to develop plants with resistance to viruses. When the replicase of the virus is expressed in transgenic plants, this also provides resistance to the virus from which the replicase was cloned. Expression of coat protein and replicase are the best examples of what is called pathogen-derived resistance, where part of the virus is used to make plants resistant to that pathogen.

Other strategies that have been used include:

1. Using antisense RNA against the viral genome.
2. Expressing an enzyme from mammals that modifies the viral RNA, inhibiting its replication
3. Expressing ribozymes, short RNA molecules that are able to bind to specific RNAs in the cell, in this case the viral RNA, and break it into smaller pieces.

All of these have been shown to work to some degree, but none have been developed into a commercial application at this time. The final method for making plants resistant to viruses targets the movement of viruses from one plant cell to another. This intercellular movement of viruses requires that the plasmodesmata, the pores between cells, be enlarged. Many viruses produce amovement protein that is responsible for this modification of plasmodesmata. From detailed studies on viruses like TMV, it was shown that some strains of virus were unable to move from cell to cell and produced very limited disease symptoms.

The reason these viruses could not move was because they have a defective movement protein, one that prevented the plasmodesmata from enlarging. When this defective movement protein was expressed in transgenic plants, it was found to provide resistance against viruses, not by preventing the infection of plant cells but by preventing the spread of virus from cell to cell.

The risks associated with the introduction of transgenic plants carrying virus-derived sequences include heterologous encapsidation, synergism and recombination. Heterologous encapsidation ('trans-encapsidation') describes the complete or partial encapsidation of the genome of one virus with the coat protein (CP) of another virus. Synergism describes a phenomenon in which one virus may complement the effects of other viruses resulting in more severe disease. Virus recombination describes the exchange of genetic material between two co-infecting viruses or between a single infecting virus and a virus-derived transgene with the potential for generating a chimeric virus with a novel pathology. Of these three, only recombination has the potential to persist in the environment in the absence of the transgenic plant.

The intention in writing this review was to provide the reader with an informative digest of our present understanding of the nature and significance of virus recombination in transgenic plants containing virus-derived sequences. To this end, the review is set-out as follows:

1. Evolutionary significance of recombination in plant viruses.
2. Mechanisms of pathogen-derived resistance (PDR) in transgenic plants.
3. Evidence of recombination among plant viruses and in plants.
4. Risks associated with recombination and transgenic plants.
5. References cited.

The review focuses almost entirely on RNA recombination. This is because the vast majority of plant viruses and virus-derived sequences (promoters and transgenes) used in transgenic plants have an RNA origin. While there is presently little published information on other types of genetic exchange involving plant viruses the following generalities can be stated:

1. Recombination among RNA plant viruses appears to be relatively frequent.
2. Recombination among DNA plant viruses appears to be relatively frequent.
3. Recombination between RNA and DNA plant viruses appears to be relatively infrequent.
4. Recombination between RNA or DNA viruses and host plant DNAs appears to be relatively infrequent.

Two types of transgenic plant are relevant to discussions on recombination (i) GM plants in which virus-derived promoters are used to control the expression of non-viral transgenes, and (ii) GM plants in which virus-derived transgenes are used to confer disease resistance in virus resistant transgenic plants (VRTPs). Virus-derived promoters are used in a wide range of plants that are not necessarily hosts of the virus (or its relatives) from which the promoter sequence was originally sourced. Since the key constraint on genetic exchange is the degree of homology (relatedness) shared by donor and recipient sequences (next section) recombination involving virus promoters will only be an issue for relatively few of the plant species in which the promoter is deployed.

Evolutionary Significance of Recombination in Plant RNA Viruses

Two distinct types of genetic exchange operate in RNA viruses. The first, re-assortment, occurs only in segmented viruses and involves the exchange of one or more discrete RNA molecules that comprise the multipartite virus genome. The second type of genetic exchange involves the introduction of a donor nucleotide sequence into a single contiguous acceptor molecule to form a novel recombinant RNA molecule and can occur in segmented as well as unsegmented viruses.

Nomenclature

Based on the similarity of the parental RNA molecules, two types of RNA recombination have been defined: homologous recombination and non-homologous recombination. Homologous recombination (HR) occurs between two similar or identical RNAs involving precisely matched (precise HR) or divergent (aberrant HR) RNA sequences. Non-homologus recombination (NHR) involves the exchange between identical viruses in which strict alignment is not maintained or involves exchange of unrelated sequences. This nomenclature is not entirely satisfactory since the term 'homologous' implies common ancestry.

The recombination mechanism

Two mechanistic models have been proposed to account for RNA recombination:

1. A breakage-rejoining model which supposes the cleavage at two distinct sites of the same or two different precursor RNAs followed by ligation of the two cleaved molecules into one recombinant RNA
2. An RNA dependent RNA polymerase (RdRp) mediated copy-choice model, also known as 'template switching', in which the viral

RdRp complex switches from one RNA template to another during replication. Where replication of the new strand continues precisely where it left the first template RNA the recombinant virus is homologous and non-homologous when it does not.

The only experimental evidence in support of the breakage-rejoining model for RNA recombination has come from Chetverin et. al. (1997) from in vitro studies of the Q? bacteriophage in which only non-homologous recombinants were observed. Nevertheless, almost all the other evidence supports the RdRp-mediated copy-choice model although it should also be noted that the template-switching mechanism has yet to be formally demonstrated.

Since template switching occurs during RNA synthesis several physical requirements need to be met for successful cross-over to occur. The first of these is that the RNA polymerase, which is responsible for catalysing the synthesis of the new RNA molecule, must first pause and then dissociate from the template viral RNA (or DNA) molecule. The second condition is that an alternative template must be sufficiently close for the dissociated polymerase to bind to. The third condition is that the new template has a polymerase binding site to facilitate cross-over and continued synthesis on the new RNA molecule.

Evidence indicates that these conditions are generally met. Transcriptional pausing has been observed in a range of RNA viruses and RNA polymerase binds only weakly to RNA, which increases its potential for dissociation. Also, in most RNA viruses, RNA synthesis occurs in membrane bound cellular compartments and this may promote the local concentration of viral RNAs making template switching by polymerase more likely. The tendency for some regions of an RNA molecule to form double stranded 'heteroduplexes' may also serve to promote both transcriptional pausing and physical bridging of the template RNAs by the polymerase. The mechanism by which polymerase recognises binding sites on an alternative RNA template probably involves either sequence complementarity (where homologous recombination occurs), or recognition of common RNA secondary structure where non-homologous recombination occurs, or both.

Evolutionary advantages of recombination

Evidence derived in the main from nucleotide sequence data has been used to compile a large and rapidly growing list of RNA viruses that have evolved as a result of genetic recombination. It is now unequivocal that recombination has played a central role in the evolution of many RNA viruses. Theoretical explanations of the evolution of sexual reproduction tend to emphasize the spread of advantageous genes and the removal of deleterious genes as the main drivers. The second of these derives from 'Mueller's ratchet' theory which predicts the gradual build-up of deleterious alleles in finite asexual populations. RNA viruses represent one of only a few examples that have been used to empirically test this prediction and results have generally indicated agreement i.e. decreased fitness in populations in the absence of reassortment.

Direct experimental evidence supporting a similar role for recombination has not been provided to date, but expectations are that genetic exchange through recombination provides for a similar function to reassortment by fixing advantageous alleles in the population and removing disadvantageous alleles by combining mutation free parts of different genomes. Although it has been proposed that recombination in monopartite RNA viruses and reassortment in segmented RNA viruses represent alternative evolutionary pathways in these two groups. The weight of evidence for plant viruses, as well as other RNA viruses, indicate that these two mechanisms are not necessarily exclusive and in some cases may even operate simultaneously.

The evidence that has been presented in favour of the theory that recombination can eliminate deleterious alleles has come from studies that show that weakly replicative mutant tombusviruses, for example, can be reconstituted through recombination with intact wild-type homologues (White and Morris 1994). Similar evidence has also been generated for the bromoviruses (Rao and Hall 1993). Several landmark studies have also shown that mutant viruses can repair their genomes through genetic recombination with virus transcripts in VRT plants carrying viral derived transgenes. Recombination between viruses and host transgenes under conditions of strong selection has been shown not only to produce viruses that cause disease symptoms quite distinct from the parental strain (Green and Allison 1994), but also hybrid viruses more pathogenic than the parental strains (Jacquemond et al. 1997). Two important studies have shown that RNA recombination in plant viruses may even provide a telomerase-like function by repair to the 3' ends of satellite RNAs, for example, of turnip crinkle virus (Burgyan and Garcia-Arenal 1998) and cucumber mosaic virus (Simon and Nagy 1996).

In addition, a growing body of information indicates that recombination between viral and host genomes has occurred (as illustrated by sequence comparison) between a luteovirus and a plant chloroplast exon. Closteroviruses have also contributed supporting evidence with sequence data indicating that this group has evolved in part by acquiring various host protein-coding genes. Evidence for the uptake of viral sequences by plant hosts is also beginning to emerge with a report of the suspected integration of Banana Streak Badnavirus (BSB) into the Musa genome. Presumably, the selective advantages to host and pathogen are resistance to or amelioration of disease and in maximising disease transmission, respectively.

Recombination among RNA viruses is probably relatively common but only hybrids for which the benefits of exchange outweigh the costs and for which the hybrid is able to compete with parental strains are likely to persist. As with mutation, genetic exchange is a random process and the vast majority of new combinations will either be non-functional or suffer decreased fitness compared to the parental viruses. As such, the vast majority of recombinants are likely to be rapidly eliminated from the virus gene pool.

Mechanisms of Pathogen-Derived Resistance (PDR) in Transgenic Plants

The incorporation of viral-derived sequences into plants has provided the means of combating economically significant crop viruses for which there are no readily accessible sources of natural resistance. Only two plant genes demonstrating resistance to viruses have been isolated to date. These include the N gene from tobacco (Erikson et. al. 1999) and the Rx gene from potato which encodes a hypersensitive response to tobacco mosaic virus (TMV) and an extreme resistance potato virus X (PVX) respectively. However, the mechanism by which each of these responses is induced is not known and while the N gene has been successfully transferred to tomato, the wider applicability of these transgenes to other crop plants has not been demonstrated.

Strategies for pathogen-derived resistance (PDR) are divided into those that require the production of proteins and those that require only the presence of RNA sequences. Generally speaking, protein-mediated strategies provide incomplete resistance to a wide range of RNA viruses, whereas RNA-mediated resistance provides very high levels of resistance to a narrower range of viruses (Beachy 1997).

Protein Mediated Resistance

Coat protein (CP) mediated resistance

Coat protein mediated resistance (CPMR) can provide either broad or narrow resistance to RNA viruses. The coat protein of TMV for instance is effective against closely related strains, but resistance decreases with diminishing sequence similarity. In contrast, the CP gene of soybean mosaic virus (SMV) which is not capable of infecting tobacco, confers resistance in tobacco to two unrelated potyviruses, potato virus Y (PVY) and tobacco etch virus (TEV). It is not known why some coat proteins afford high levels of protection and why others provide broader or lower levels of resistance, though this issue will almost certainly have direct relevance to the risks of virus recombination where transgenes from several virus strains are required to deliver sufficiently broad resistance.

Replicase - mediated resistance

Genes encoding viral replicase proteins can confer near immunity to infection. All plant viruses have replicase genes and transformation of plants with both defective and non-defective replicase genes can provide very strong resistance - though usually only to the donor virus or very closely related strains. The mechanisms involved in replicase-MR are not known but are likely to involve the suppression of virus replication and gene expression. While a feature of replicase-MR is that it is generally quite specific, some interesting exceptions have been observed. In one case, transgenic resistance to potato leaf roll virus (PLRV) turned out to be effective against a wide range of PLRV strains. In another example, the TMV replicase gene containing an Agrobacterium sequence insertion was found to confer high level resistance in tobacco plants to a wide range of tobamoviruses.

Movement protein - mediated resistance

Movement protein mediated resistance (MPMR) is thought to arise through competition between the dysfunctional introduced MP and the invading virus for binding sites on host plasmodesmata resulting in reduced local and systemic spread of infection in MPMR transgenic plants. In contrast to replicase-MR, the expression of dysfunctional MP genes facilitates resistance to a much wider range of viruses. This approach has not seen wide application due to the scale of effort necessary to make and screen banks of mutant MP genes to identify effective candidate mutants.

The expression of functional MPs appears either to have no effect or actually increases susceptibility of plants to virus disease. It has been anticipated that improved knowledge of MP structure and function will lead to the development of mutant MPs that can act as highly effective inhibitors of many different types of virus (Beachy 1997).

RNA-mediated resistance

Several PDR strategies involve the expression of genes that do not encode proteins. One of the first was expression of antisense RNA sequences to inhibit virus replication (Hammond and Kamo 1995) although some doubt exists as to whether this effect was actually due to RNA suppression. Other mechanisms may also be responsible including interruption of template selection by viral replicase and/or the formation and subsequent degradation of double-stranded RNA. Antisense mediated suppression is generally very narrow in terms of the range of viruses that it targets.

RNA mediated resistance in transgenic plants has also been shown to operate through competitive inhibition of the invading virus with the transgene or its RNA transcript acting as a decoy for proteins needed for virus replication and movement. This process is known to operate in instances where the transgene specifies a 'defective-interfering' (DI) viral RNA as, for example, in transgenically expressed turnip yellow mosaic virus (TYMV) 3' DI RNA which competes with the invading TYMV genome (Zaccomer et.al. 1993). Satellite RNAs are similar to DI molecules but do not depend on sequence similarity with the donor virus.

Transgenic plants encoding DI RNAs (or DNAs) and satellite RNAs have been tested for their capacity to reduce replication and ameliorate disease with some degree of success. Presently, the most well known type of RNA mediated resistance is RNA suppression - a process that describes the post-transcriptional destruction of viral RNA also known as "gene silencing" (Baulcombe 1996). Virus gene silencing was suspected early on in the development of VRT plants from observations that in plants designed to be resistant through CP expression; protection was maintained even in the absence of the transprotein (Lindbo et. al. 1993). Gene silencing in plants derives either from a block on the initiation of viral transcription or from degradation of viral RNA after transcription and is described as transcriptional gene silencing (TGS) and post-transcriptional gene silencing (PTGS) respectively. In either case, the host cell activates a destruction mechanism upon detecting aberrant RNA sequences and both viral and transgene RNA is destroyed.

Suppression of RNA Silencing

Suppression of RNA silencing probably represents an evolutionary adaptation by plant viruses to a generic host antiviral defence strategy since many of the suppressor proteins have previously been shown to be important determinants of virulence. In fact, the ability to suppress PTGS may be crucial for many viruses to be able to infect their plant hosts. It seems that suppression occurs in a general manner, independent of the gene or sequence responsible for inducing the silencing mechanism that, by contrast, can only act against near homologous sequences.

Transgenic plants designed to work by gene silencing are only effective because the PGTS mechanism is activated prior to infection and is therefore able to degrade the target RNA before viral proteins can be translated from it.

Evidence of Recombination among Plant RNA Viruses and Transgenic Plants

Mutation, reassortment and recombination are the drivers for evolution in RNA viruses. Recombination probably plays an additional role in rescuing viruses by repairing mutations in virus genomes that have occurred as a result of proof-reading errors during replication. The realisation that recombination has a central role in the evolution of many RNA viruses began with experiments on poliovirus by Hirst (1962) and Ledinko (1963). Since that time recombination has been documented for many animal and plant RNA viruses. The rapid accumulation of evidence for recombination is all the more impressive when one realises that prior to these formative studies, recombination was not thought to be a property of RNA genomes at all.

Recombination in plant RNA viruses was first demonstrated for brome mosaic virus (BMV) and there is now an extensive literature documenting recombination among both plant and animal RNA viruses as well as in bacteriophages and double-stranded and negative sense RNA viruses. The genetic engineering revolution, which has brought with it the development of the polymerase chain reaction (PCR), automated nucleotide sequencing techniques and phylogenetic analysis, has transformed the study of virus recombination by providing the means of detecting and characterising recombination events. Using these techniques it has been possible not only to detect rare recombination events and exchanges that have occurred long ago but also to identify the precise location of the cross-over point. Two types of data have been used to provide evidence of RNA recombination, one based on phylogenetic analysis and the other based on empirical observation.

Detecting Recombination among RNA Viruses

Methods for analysing putative recombination events are often graphical in nature and exploit the fact that many recombinant sequences are mosaics of virus phylogenies. For example, 'Split Decomposition' analysis employs a split graph highlighting transition from a dichotomously branching tree form indicative of sequence relationships to a complicated network indicative of a history of genetic exchange. Other graphical programmes look for discordant nucleotide sequence relationships suggestive of recombination, examples of which include 'boot scanning', 'PhylPro', 'TOPAL' and 'DIVERT'.

Statistical approaches employ procedures based on maximum-likelihood computation of goodness-of-fit estimates by comparing the frequency distribution of polymorphisms in putative parental and recombinant strains. Some statistical approaches attempt to quantify the amount of recombination rather than document specific recombination events and do this by calculating an 'Index of Association' which measures the degree of linkage between alleles at different loci. A further development on this approach is the 'Homoplasy Test' which compares the number of convergent and parallel base changes (homoplasies) in a maximum parsimony tree to the number expected by chance alone where excessive homoplasy indicates a history of recombination.

There is also an abundance of direct experimental data demonstrating homologous RNA recombination using movement-defective viruses in which functional recombinants have been restored through genetic exchange. Non-homologous recombination (NHR) between plant viruses has been demonstrated in two instances. The first involved co-inoculation of protoplasts and plants with defective cucumber necrosis virus (CNV) and tomato bushy stunt virus (TBSV) RNAs which produced recombinant defective interfering (DI) strains as well as fully infectious derivatives. The second case involved the co-inoculation of tomato aspermy virus (TAV) RNAs 1 and 2 and cucumber mosaic virus (CaMV) RNAs 2 and 3, resulting in a novel cucumovirus encoding a novel replicase comprising TAV helicase and CaMV polymerase subunits.

The phylogenetic evidence as well as the empirical evidence indicates that not only do different RNA viruses not recombine at the same frequency, but also that some strains are highly recombinative while others show no evidence of recombination at all. Phylogenetic analyses of potyvirus strains, for example, indicate numerous recombination events, whereas analyses of cucumovirus genomes have yielded no evidence at all have past recombination events. While these findings may be explained to some extent by differences in the detection limits of the different virus systems, it is also likely to be due, in part at least, to the inherent characteristics of the virus-host interaction. This last point, in particular, is significant for VRT plant risk assessment since almost nothing is known about the role of host factors in viral RNA recombination.

Recombination in VRT Plants

Numerous transgenic crops have been developed for tolerance or complete resistance to a range of economically important virus diseases. Recombination in transgenic plants has been investigated in systems in which movement defective viruses are inoculated into plants expressing the functional homologous movement protein gene, and then tested for the presence of systemic (recombinant) viruses. This phenomenon has been reported in three experimental systems: cauliflower mosaic virus (CaMV), cowpea chlorotic mottle bromovirus (CCMV) and tomato bushy stunt tombusvirus (TBSV).

The first extensively documented case of recombination involving a transgene was in Brassica napus expressing the cauliflower mosaic virus (CaMV) translational transactivator gene ORF VI. Experimental plants were inoculated with a mutant CaMV strain lacking the ORF VI gene and recombination with the transgene restored systemic movement to the inoculated strain.

In an important related study a CaMV strain unable to infect solanaceous plants acquired ORF VI from transgenic tobacco, Nicotiana bigelovii, generating recombinants with altered symptoms and extended host range. A strain of cowpea chlorotic mottle bromovirus (CCMV) that had been crippled by deletion of part of the coat protein gene was used to elegantly demonstrate recombination in transgenic N. benthamiana containing the missing CP sequence. In this example, mutation markers unequivocally confirmed a single recombination event between the transgene mRNA and the movement-defective CCMV to generate fully restored systemic hybrids containing the functional CP gene. Sequence analysis of the recombinant strains revealed numerous modifications flanking the putative cross-over site, and when tested on a range of host species, four out of seven recombinants displayed novel symptoms, although none was more fit than the wild-type virus.

In these experiments, recombinant viruses were recovered under conditions of strong selection pressure in that the challenging virus was movement defective and had to recombine with the transgene in order to regain function. Searching for recombinant viruses generated under conditions of little or no selection pressure is generally regarded as necessary for realistic risk assessment.

Perceived Risks Associated with Virus Recombination and Transgenic Plants

The key characteristic that distinguishes genetically modified organisms (GMOs) from most other technologies (e.g. chemical pesticides) is that GMOs can replicate and, consequently, once established in the environment a GMO may be able to persist indefinitely and prove difficult and costly to remove if this becomes necessary. Persistence does not just imply temporal continuity but also indicates spatial spread since not only could the entire GMO disperse beyond the boundary of the intended release site but the transgene itself may be able to disperse into the wider environment as a result of out-crossing between the GM plant and compatible wild relatives.

Uptake of the transgene by wild-type organisms provides the potential for unique and unintended combinations of genes and for this reason considerable attention is now being directed at understanding the risks associated with gene introgression. In addition to out-crossing, VRT plants present an additional concern because of the possibility that the viral derived transgenes used to confer disease resistance in GM crops may recombine with the genomes of naturally occurring plant viruses.

Ecological Risk Experiments

There are few published ecological studies assessing the potential for environmental impact from the cultivation of virus recombination in VRT plants. Agriculturally important plant viruses are capable of impacting on natural host plant populations and that impacts are unlikely to be uniformly distributed among host populations. This was almost certainly as a result of differences in the distribution and behaviour of the aphid, weevil and flea vectors responsible for transmitting these virus pathotypes. Since evidence from the controlled trials discounted the possibility that vector species had imposed differential selection for virus resistance among the study populations, the conclusion was that the potential for virus induced impacts on wild B. oleracea populations will be highly dependant on the distribution and behaviour of the vector species involved.

Viral Diseases of Maize

The most important viral diseases of maize are maize dwarf mosaic (MDM) and maize chlorotic dwarf (MCD). Yield losses in fields where the disease is well established may be severe depending on the susceptibility of the corn hybrid being grown. In general, most loss occurs when the plants become infected at the "knee high" stage of development and is minimal on most hybrids if infection occurs after tasseling. Diagnosis of virus infected plants is difficult based on field symptoms. Samples of plants should be tested in the laboratory to confirm the presence of the virus.

Maize dwarf mosaic (Symptoms)

1. Maize dwarf mosaic symptoms vary with the corn hybrid and the stage of development of the plant at the time of infection.
2. Early infection results in chlorotic spots or flecks that elongate in young leaves in the whorl. Flecks merge into chlorotic streaks along the leaves.
3. These streaks form mosaic or mottled patterns and may turn to a general yellowing as the growing season progresses.
4. Later, plants may have blotches or streaks of red that generally appear after periods of cool (60°F) night temperatures.
5. Infected plants are predisposed to root rot and may be barren.

Disease cycle

Maize dwarf mosaic virus (MDMV) exists in several strains, the most common being strain A, which infects and overwinters primarily on johnsongrass. Strain B does not infect johnsongrass. Besides corn and johnsongrass, the MDM virus may infect over 100 wild and cultivated grasses. More than 20 species of aphids can transmit MDMV. An aphid can acquire the virus within a few minutes of feeding on an infected corn or johnsongrass plant. The aphid then flies or is carried by the wind to other corn plants and inoculates them with the virus. The aphid retains, and is generally able to transmit the virus for 15-30 minutes after acquiring it. The corn leaf aphid and the green peach aphid are common aphid vectors of MDMV.

Maize chlorotic dwarf (Symptoms)

1. Maize chlorotic dwarf symptoms include yellowing of youngest leaves in the whorl and a distinct fine yellow striping, or vein clearing of the smallest veins visible between the larger veins.
2. This chlorotic vein clearing of secondary veins is diagnostic for maize chlorotic dwarf and may be more readily observed on the undersides of infected leaves.
3. The striping may not be seen in later stages of development due to leaf reddening and general yellowing of the plant.
4. Affected plants may also be stunted due to shortening of the upper internodes.

Disease cycle

Maize chlorotic dwarf virus (MCDV) is primarily transmitted by the black-faced leafhopper. The leafhopper must feed on an infected plant for several hours to acquire MCDV and then can transmit it for up to 48 hours.

Control Measures

1. Grow hybrids tolerant or resistant to maize dwarf mosaic. There is good tolerance and resistance to strain A, but only fair tolerance and no resistance to strain B in dent corn. There is no resistance to maize chlorotic dwarf virus and only fair tolerance.
2. Destroy grasses hosts, including volunteer corn in areas where corn is to be planted.
3. Plant early, since early-planted corn will escape damage because aphid populations do not build up until the plants are past the seedling stage.

Viral Diseases of Peanut

Unlike many of the pathogens of peanut that depend on the forces of wind or rain for dissemination, a vector actively disseminates most viruses. Insects such as aphids or thrips are common vectors, but people and equipment may also spread viruses by mechanical damage of plants. Yield losses from virus diseases in peanut depend on the severity of the disease, the percentage of plants infected, and the plant age when infection occurs. Yield loss is generally high if a large number of plants are severely infected early in the growing season. Yield loss is usually minimal when infection occurs late in the season when, most or all of the pods are mature, or by mild viruses, regardless of when plants are infected.

Peanut mottle (Symptoms)

1. The symptoms of PMV infection can vary with cultivar, time of infection and environment. The most common symptom, although it may not be readily noticed, is a mild mottle or mosaic on the youngest leaves of infected plants.
2. The light and dark green areas of affected leaves can best be seen if leaves are held up to light.
3. Margins of leaflets may curl up and depressions in the leaf tissue between the veins may become prominent. Plants are generally only mildly stunted, if at all.
4. As plants mature, the symptom expression generally declines, particularly during hot and dry weather.

5. Pods from infected plants may be reduced in size and have irregular gray to brown patches. The seed coat of affected seed may also be discolored.

Disease cycle

Peanut seed is thought to be the most important primary source of PMV for the beginning of the disease cycle. Commercial seed lots usually have a low frequency of PMV infection (less than 1 percent). However, an infection level of 0.1 percent can result in about two infected seedlings per 100 square yards in a field. Aphids may then transmit the virus from infected to healthy seedlings. At least six species of aphids mechanically transmit PMV. Winter forage legumes such as clovers are also hosts for PMV and may serve as sources for the virus. can efficiently transmit the virus. Tractor tires crushing infected vines in touch with healthy vines may also.

Control measures

1. Because the virus generally has a minor effect on commercial peanut yields, no specific management practices are recommended for control of PMV.
2. Several breeding lines with resistance have been identified, but this resistance has not been incorporated into any commercial varieties.
3. The use of PMV-free seed is the most feasible approach for control, as this prevents the disease from becoming initially established in the field. PMV-free seed can be produced under rigorous procedures. If PMV-free seed is used, volunteer plants must be completely removed and the field situated so that PMV hosts, such as clovers, southern pea, and navy bean are at least 100 yards away.

Peanut Stunt (Symptoms)

1. Peanut Stunt virus(PSV) causes striking and destructive symptoms on peanut. The most common symptom of PSV infection is severe stunting of the plant.
2. Stunting may affect the entire plant or a portion of the plant, such as a single branch. 3. Infection early in the season generally results in the most severe expression of symptoms.
4. Leaves of infected plants may be curled upward and be malformed. Leaves may also appear yellowed, mottled, or have a dark green banding pattern.
5. Pod production is reduced in both size and number. Pods may also be malformed and hulls may split.

6. Seed germination and seedling vigor is reduced when seed from infected plants.

Disease cycle:

PSV is introduced into fields either by infected seed or by aphids, which acquire the virus from another plant host plant, such as white clover. PSV infects at least 115 species of plants. Important economic hosts include peanut, bean, southern pea, and white clover. The virus can be spread mechanically, by aphids, and in seed. Transmission in seed is not thought to be of great significance, because the percentage of transmission from seed is low (less than 0.3 percent) and plant viability from infected seed is low. Aphids, however, play an important role in spread of the virus. Three species of aphids are known to transmit PSV. Aphids may acquire the virus from infected hosts such as white clover or weeds such as crownvetch. The virus may then be carried to nearby peanut fields.

Control measures

1. Peanuts should be planted in fields free of PSV and away from known reservoirs of PSV, such as white clover.
2. The use of peanut seed free of PSV is also recommended.
3. Infected plants should be rogued from seed fields.
4. Cultivars with resistance to PSV have not been developed, although some breeding lines with resistance have been identified.

Tomato spotted wilt (Symptoms)

1. Symptoms of TSWV in peanut can be quite variable. A common symptom of the virus in peanut is the development of yellow to pale green ringspots of various sizes and shapes on infected leaves.
2. These are best seen when infected leaves are held up to light.
3. Stunting also occurs and can be quite severe if plants are infected early in their growth.
4. Infected plants may produce distorted, mottled, and smaller than normal leaflets.
5. Plants infected within 45 days of planting produce few, if any, marketable pods. Yield losses decrease when as progressively older plants are infected. Late-season infections usually have little impact on yield.
6. Death of terminal buds may occur and plants may be killed in extreme cases.
7. Dead plants are usually found in fields with a high incidence of plants that were infected early.

8. Seed produced on infected plants may be reduced in size, malformed, and have discolored seed coats.

Disease cycle

TSWV is not transmitted to plants through infected peanut seed. TSWV infects more than 200 species of plants. These plants include economic crops such as tomato and pepper as well as weed species. Winter annuals such as chickweed, shepardspurse, swinecress, and sowthistle are known to be TSWV hosts. Weeds may serve as a reservoir of the virus in areas affected by tomato spotted wilt. The virus is spread by at least five species of thrips. Thrips become infective only when they acquire the virus while feeding on infected plants as larvae. Larvae can transmit the virus before they pupate, but more commonly adults transmit the virus. Adults remain infective throughout their life. Thrips are generally a widespread insect pest of peanut. Therefore, the potential exists for TSWV to be an important disease in peanut each season.

Control measures

1. Thrips control with granular insecticides is usually recommended where losses from TSWV have occurred in peanut. The success of thrips control in managing TSWV has been variable. This is perhaps because these insecticides are systemic and infective thrips are able to feed for short periods and infect plants before the ingested insecticide acts to kill them.
2. Maintenance of weed control in areas adjacent to peanut fields may help reduce potential reservoirs of the virus.

Viral Diseases of Bean

Bean common mosaic virus (BCMV): Symptoms

1. Systemically infected plants, especially those from infected seeds, have leaves with green mosaic patterns and distortions (curling, strapping, or puckering of tissues along leaf veins).
2. Plants may be stunted and have only a few pods which mature later than uninfected pods.
3. Vascular tissue can become necrotic, producing dark streaks on petioles and stems.
4. At high temperatures (above 78 F), cultivars with the hypersensitive resistance gene (I gene) respond to necrosis-inducing strains of BCMV with a systemic necrosis called black root. Plants with black root die.

5. Nonsystemic infections can appear as ring-like lesions on foliage.
6. Appearance and severity of symptoms depends on strain of the virus, variety, time of infection, and environmental conditions.

Disease cycle

BCMV is a seedborne virus transmitted by aphids and mechanically by plant sap. Many types of beans, alfalfa, and common clover are hosts.

Control measures

1. More than 15 strains of BCMV are known and breeders have incorporated resistance to the more important strains in many commercial varieties.
2. Certified seed programs are restrictive for BCMV contamination.
3. Controlling large populations of aphids can reduce spread.
4. The primary control is selection of high quality, virus-tested seed of genetically resistant varieties.

Bean yellow mosaic virus (BYMV): Symptoms

1. It causes plant stunting and leaves with contrasting areas of dark green and yellowed tissue.
2. Bright yellow spots can be obvious on older plants.

Disease cycle

This virus is readily transmitted by aphids and mechanically by plant sap, but it is not seedborne. The virus also attacks wild hosts such as clover and sweet clover. Some viral and fungal diseases of broad-acre crops can only survive from the end of one winter growing season to the start of the next on green plant material - the so-called 'green bridge'. The aphid vectors of many crop viruses also require this 'green bridge' to survive between crop growing seasons.

Control measures

1. There is no specific management practices are recommended for control of BYMV.
2. Genes for resistance have been identified.

Viral diseases of clovers (Symptoms)

1. Virus symptoms on clovers vary somewhat, depending on host species and the virus involved.
2. In general, viral infected plants have mild to severe leaf mottling.
3. Narrow, pale to yellow discolored areas may be found along the veins, or large, light green to yellowish blotches may occur between the leaf veins.

4. In some cases, leavescurl, or are puckered or ruffled.
5. Severely affected plants may be dwarfed or weakened and unable to withstand prolonged drought or severe winters.
6. The weakening caused by viral infection may predispose plants to attack by other disease causing agents.
7. Symptoms of most clover virus diseases are conspicuous during cooler periods, disappearing temporarily during hot weather.

Some examples of viral diseases of commonly grown clovers:

(a) *Red clover*. Red clover vein mosaic is one of the most prevalent and widely distributed viruses of this host. In the field, red clover also is a natural host for yellow bean mosaic and potato yellow dwarf viruses. The viruses of alfalfa mosaic, pea common mosaic, and white clover mosaic also can infect red clover.

(b) *Alsike clover*. Virus mosaic of alsike clover is widely distributed. In addition, alsike clover is known to harbor viruses of annual legumes, principally those of peas.

(c) *White clover* (including Ladino clover). In the field, white and Ladino clovers display at least two kinds of viral disease symptoms. Some plants develop vein clearing with mild mottling while others develop bright yellow patches or streaks between the veins of leaves. The vein clearing and mottling symptoms are commonly caused by white clover mosaic, which consists of a mixture of two viruses, pea mottle and pea wilt. Yellow blotches between the veins, particularly noticeable in Ladino clover, are caused by a strain of alfalfa mosaic virus. A similar strain of virus from alfalfa and Ladino clover causes a severe tuber necrosis in potatoes.

(d) *Sweet clover*. Many viruses can infect sweet clover in the field. For example, the following have been recovered from infected plants: alfalfa mosaic, yellow bean mosaic, several pea viruses, red clover vein mosaic, white clover mosaic, tobacco streak virus, and the Colorado rednode virus of bean.

Disease cycle

The viruses that attack clovers can also infect many related and unrelated hosts. In turn, some viruses from hosts such as peas, beans, potatoes, alfalfa, and certain weeds are readily transmitted to clovers. Thus, a virus-infected crop in one field or plants along a fencerow may act as the reservoir for virus infection of an entirely different crop in a neighboring field. The spread and severity of infection depend on insect populations carrying viruses from plant to plant, the relative susceptibility of the host to the viruses present, and the age of the plants when infected.

Control measures

1. To date, very little has been done to control viral diseases of clover.
2. Where the crop is being grown for seed, insecticides may be applied to reduce the number of insect carriers; clover viruses are rarely seedborne. When possible, clover fields should not be planted adjacent to other leguminous crops such as peas or beans.
3. The ultimate solution is the development of varieties of clovers resistant to the most prevalent and damaging viral diseases.

Viral Diseases of Soybean

Soybean mosaic (Symptoms)

1. Symptoms of SMV vary depending on the soybean cultivar, the age of the soybeans, the virus strain, and the temperature.
2. Symptoms are most noticeable under cool temperatures of 18 to 24^{0}C. When temperatures rise above 30^{0}C, leaf symptoms may be masked.
3. The youngest and most rapidly growing leaves show the most severe symptoms. The leaves of SMV infected plants are distorted and narrower than normal, and develop dark green swellings along the veins.
4. Infected leaflets are puckered and curl down at the margin.
5. Plants infected early in the season are stunted, with shortened petioles and internodes.
6. Diseased seed pods are often smaller, flattened, less pubescence, and curved more acutely than pods of healthy plants.
7. Infected seed are mottled brown or black, usually smaller than seeds from healthy plants, and germination may be reduced.

Disease cycle

Soybean mosaic is caused by soybean mosaic virus (SMV) and is the most widely distributed virus diseases of soybeans. It is spread by planting diseased seed and by at least 31 species of aphids

Bud blight of Soybean (Symptoms)

1. Infected plants are stunted and the pith of stems and branches show a brown discoloration, first near the nodes, then throughout the stem.

2. Leaves on infected plants are smaller, wrinkled, and have a bronze discoloration.
3. Buds become brown, necrotic, and brittle; hence the name bud blight.
4. The most strirking symptom is the curving of the terminal bud to form a crook.
5. Diseased pods are often aborted or contain no seed. In the field, plants that are infected remain green after healthy plants have matured.

Disease cycle

Bud blight, caused by the tobacco ringspot virus (TRSV), can be a serious disease of soybeans. Yields may be reduced 25-100% depending on the time of infection. Planting infected seed spreads the virus, but the amount of infected seed produced is usually extremely low.

Bean pod mottle (Symptoms)

1. Plants infected with both soybean mosaic and pod mottle are stunted, have distorted foliage, misshapen fruit, and necrotic tissue.
2. Seed from plants infected with pod mottle are smaller than normal.

Disease cycle

This disease, caused by the bean pod mottle virus (BPMV), often occurs in combination with soybean mosaic virus. Mainly the feeding of certain insects, particularly the bean leaf beetle, transmits the virus. The virus is not seed transmitted. In the field, diseased plants show a mild yellow mottling on young actively growing leaves. As these leaves approach maturity the mottling becomes masked.

Peanut mottle (Symptoms)

1. A mosaic of dark green and yellow areas is produced in older leaves.
2. Ring patterns of yellow patches develop on the third and fourth leaves following infection. Infected leaves pucker and curl downward at the margins.

Disease cycle

Peanut mottle is caused by the peanut mottle virus (PMV). It was first observed in 1971 and is the most common virus disease of soybeans in Georgia. The virus is seedborne in peanuts, but not in soybeans. Aphids can carry the virus from infected peanut fields or from other legumes in nearby fields. Volunteer peanut plants remaining in soybean fields are a source of inoculum for soybean infection.

Viral Diseases of Wheat

Yellow dwarf disease (Symptoms)

1. Leaf discoloration is usually the most notable early season symptom.
2. Leaves may be various shades of red to purple or pinkish-yellow to brown.
3. As an infected plant continues to grow, older leaves typically begin to die back from the tip and may feel somewhat leathery, while the new leaves begin to discolor.
4. Spring infections occur as well, but these commonly discolor only the flag leaf and do not cause significant yield reductions.

The Virus

The virus particles (virions) of Barley Yellow Dwarf Virus Disease (BYDV) are concentrated in phloem cells. They are polyhedral in shape, 21 to 26 nm in diameter. Isolates of BYDV differ serologically and in virulence, host range and vector specificity. Most strains are compatible in wheat and aphids, but some are mutually exclusive, having different compatibilities with certain aphid species. Seed, soil, sap or other insects do not transmit BYDV.

Transmission

BYDV is transmitted by more than 20 aphid species. Corn leaf aphids, English grain aphids, greenbugs and oat bird cherry aphids are among the most common vectors. Aphids acquire the virus by feeding on diseased plants for periods as short as 30 minutes but usually 12 to 30 hours. After a four-day latent period, the aphids can transmit the virus to healthy plants when feeding. After acquisition, aphids are able to transmit the virus as long as they live known as "persistent" or "circulative" virus. Symptom expression after feeding may be in one to three weeks, but the symptoms are very inconspicuous in fall infections.

There are four prominent strains of BYDV that are correlated with the particular vector:

1. *RMV* - transmitted regularly by Rhopalo-siphum maidis (corn leaf aphid)
2. *RPV* - transmitted regularly by R. padi (oat bird cherry aphid).
3. *MAV* - transmitted regularly by Macrosiphum avenae (English grain aphid)
4. *PAV* - transmitted regularly by R. padi and M. avenae.

Disease cycle

BYDV persists in cereal crops, in annual and perennial grasses, both tame and wild, and in the aphid vectors. Spread is dependent upon vector movement. In the fall, aphids move from grasses or volunteer cereals to newly planted wheat (or other winter cereals). In the spring, aphids that overwinter as adults in grasses or winter cereals become active vectors. Others that develop from eggs acquire BYDV in the spring by feeding on infected grasses and cereals during migration. Most aphids have a "winged model" (alate) form that provides more rapid movement. In feeding, they settle down to a non-wing form, with somewhat slower movement from plant to plant.

Aphid flights can be localized, or they can be disseminated for hundreds of miles when assisted by wind. The movement is associated with the development of the disease from south to north in North America. Missouri is a strategic state in the dissemination of infectious aphid vectors.

The fall infections are the most serious in wheat. Inoculated plants become systemically infected and develop symptoms within two weeks at 20^{0}C, within four weeks at 25^{0}C, but not at all above 30^{0}C. In other words, cool (not cold) temperatures accentuate symptom expression.

Control measures

1. Although there are no high levels of resistance in wheat, there are some tolerant wheat varieties. This is also true of red leaf tolerance in oats.
2. The use of insecticides in the fall for control of aphids in the field, if properly timed, can reduce the incidence of BYD and increase yields, especially in respect to fall infections. The economics and efficacy of insecticides for vector control must be considered. It must be emphasized that aphids can inoculate the virus within a few hours, so the timing of the insecticide, especially with the short-duration insecticides, would have to be pinpointed better than most farmers could accomplish.
3. Fall infections can be offset from periods of high aphid activity by planting later in the fall when cooler temperatures slow aphid movements.

Wheat yellow mosaic

Wheat yellow mosaic (WYM) is also known as wheat spindle streak mosaic (WSSM). There are some similarities to wheat soil-borne mosaic in respect to symptoms and dissemination. However, the wheat spindle streak mosaic virus (WSSMV) tends to be uniformly distributed in fields rather than in pockets. Symptoms of WYM are most evident on lower leaves in early spring, and like WSMB when temperatures rise, the symptoms become masked.

Symptoms

1. Symptoms appear in early spring as yellow-green mottling, dashes and streaks on leaves.
2. Discontinuous streaks run parallel with veins and taper to form chlorotic spindles.
3. When temperatures are cool, the streaks progress to the flag leaves. Discolored areas tend to coalesce, followed by necrosis.
4. Reddish streaking at the leaf tips often precedes necrosis.
5. Some stunting and poor tillering occur in infected plants.
6. Head numbers and seed production are reduced, but kernel weight appears to be maintained.
7. The virus can cause production losses if it becomes extensive.
8. The optimum temperature for symptom development is 60°F. Above 68°F disease progress stops.
9. Symptoms typically appear in early spring right after green-up. By the time of jointing, mosaic symptoms have usually faded, but stunting may persist until maturity.

The Virus

Particles (virions) of WSSMV are rods that are normally 15 to 18 nm wide and 200 to 2,000 nm in length. The threadlike virions are rather unique and they are often clustered together, characteristics that are helpful diagnostic features for electron microscopy. The WSSMV particles usually are found in epidermal and parenchyma tissues of infected leaves. They are sparse and are difficult to isolate in leaf-dip preparations.

The Vector

Interestingly, the wheat spindle streak mosaic virus has the same fungus vector as the wheat soil-borne mosaic virus. Polymyxa graminis harbors the virus, and the virus enters the wheat plants through root entry by the fungus.

Disease cycle

The virus survives for many years in agricultural soils even in the absence of wheat. Apparently, WSSMV can exist for a long time in association with the Polymyxa graminis fungus. The virus enters wheat plant roots via the fungus in the fall. Fall infections are most serious and account for the symptoms seen in early spring. Spring inoculations occur but are not as important. The disease is most significant under cool spring temperatures and is of little consequence under warm conditions. This relationship to cool temperatures probably has much to do with prevalence in any given season and also explains the association between occurrences and geographic locations. For example, there are more incidents of the spindle streak mosaic disease in southeast Missouri than in northern Missouri, where soil-borne mosaic seems to be more prevalent.

Control measures

1. Crop rotations also appear to limit incidence, in spite of the fact that the virus can survive for many years.
2. There are no pesticides, which provide economic control of soilborne mosaic. Liberal use of nitrogenous fertilizers such as urea and poultry manure decrease soil infectivity.
3. Any field with a history of WSBMV should be planted to varieties with resistance. Many varieties are resistant to WSBMV.
4. Late planting is sometimes effective in avoiding infection periods in the fall. Wheat planted after the "fly- free" date is less likely to be attacked by soilborne mosaic as well as other viral diseases.

Wheat streak mosaic (Symptoms)

1. WSM symptoms typically appear in April or May as yellow stunted areas along field margins. These areas are often associated with volunteer wheat from the former crop.
2. WSM symptoms become more serious with time, in contrast to the soil-borne viral diseases and the disease is much more destructive than barley yellow dwarf.
3. The yellowing and stunting become more severe and a gradual spread across the field may be seen.
4. The earliest emerging wheat, often volunteer wheat, usually is most affected.
5. Individual plants turn yellow, show definite stunting and may have wilting symptoms.
6. Tillers may be partially prostrate. Root development is often reduced.
7. Mosaic symptoms begin in younger leaves as light green to yellow dashes that enlarge to give a streaked appearance. Finally, the whole leaf blade turns yellow.
8. The younger the plants are when infected with the virus, the more severe the symptoms. Head formation can be nullified.

9. Late infections cause light and dark green streaks on the flag leaves, but not much stunting.
10. WSM goes to spring wheat, barley, corn rye, oats and a number of annual and perennial grasses (wild and tame).

The Virus

WSM is caused by the wheat streak mosaic virus (WSMV), which is carried from diseased to healthy plants by the microscopic wheat curl mite, Aceria tulipae. Particles (virions) of WSMV occur in most leaf cells as flexuous rods, 15 x 700 nm. The virus is relatively unstable and difficult to purify. It is easily transmitted through sap as well as the mites.

The Vector

The wheat curl mite, Aceria tulipae Keif, is the principal means of dissemination of WSMV in nature. There seems to be no evidence of other mites or insects that serve as vectors. Curl mites are extremely small, less than 1/100 inch in length, requiring a dissecting microscope or a 10 to 20 power hand lens for observation. The life cycle, from egg through two larval stages to the adult and egg, can be completed in seven to 10 days. Therefore, populations can build up dramatically. Under ideal conditions, one adult could theoretically produce several million offspring in 60 days. However, predators or other environmental conditions usually limit population buildups. Curl mite larvae acquire WSMV within a few minutes of feeding on infected plants. They are able to transmit the virus to healthy plants for at least a week following acquisition.

WSMV is carried in the mid and hind gut of all larvae and adult mite stages. The virus does not pass through eggs. The white and cylindrical mites possess four tiny legs next to the head that provide limited movement. They also have a tail end (anal) sucker for attachment. Their dispersal plant-to-plant and/or field-to-field is via wind. The mites spin webs that act as "sails." After landing on a suitable host, the mites crawl up to the youngest unrolled leaves in the whorl.

They attach themselves to the upper surface by the anal sucker and begin to feed and reproduce. Feeding prevents the leaves from unrolling normally, causing rolled leaf edges. Consequently they are known as "wheat curl mites." Heavy populations may cause enough leaf rolling so that expanding leaves and heads may be trapped. This may help to diagnose wheat curl mite infestations. As the wheat plants mature, the mites migrate upward, including the ripening heads. If grain is shattered before harvest, especially as a result of hail, the mites attached to such kernels can sometimes survive long enough to move to the sprouting seedling. In such cases, new volunteer seedlings are infected early and can be an early source for field infections. Without early emerging volunteer wheat, the wheat curl mites and the virus will die quickly following wheat maturity. Hot, dry July weather reduces mites and virus drastically. When deprived of food and water, they survive only for hours or at most two days. However, there are always mites that survive on grasses. Wet summer weather conditions permit volunteer wheat to grow, and the earlier it emerges and the curl mites become established, the higher the population. These mites can then be blown into fall seeded wheat. Warm weather in October and November extends the period of mite activity, with consequent virus inoculations. There is also some evidence that warm weather in February and March may result in further activity of overwintering mites.

Host plants

Wheat is the preferred host for curls mites and the wheat streak mosaic virus. Other cereal grains (barley, oats, triticale, rye) are slightly susceptible. Sorghum species are immune to WSMV and are poor hosts to the mites. A few corn hybrids are susceptible to the virus, and mites migrate to corn fields from mature wheat fields. Infections of corn usually do not cause serious losses, although virus symptoms can be identified. The movement of viruliferous mites from mature corn back to wheat usually is not a serious problem. The corn matures early enough that mites die before fall seeded wheat emerges. Certain summer annual grasses such as barnyardgrass, crabgrass, foxtails, sandbur and witchgrass are susceptible to the virus and can serve as potential oversummering hosts. Buffalograss, smoothbrome and western wheatgrass are perennial grasses that are immune to the virus but can support mite populations. All soft red winter wheat and hard red winter wheat varieties are more or less susceptible to WSMV. Therefore, various cultural practices must be used to minimize infection.

Disease cycle

The virus (WSMV) and the wheat curl mites persist on wheat, corn, millet and certain susceptible grasses, such as buffalograss, crabgrass and foxtail. From spring through fall, winds distribute the mites to new cereal and grass hosts. Since the virus and the mites must persist on living susceptible hosts, they are subject to extinction when wheat is ripe and gone to harvest. In mature or recently harvested fields, the mites will survive on green shoots or on volunteer plants

Table 2.1. Grass plants tested for suitability as hosts for wheat curl mite survival and wheat streak mosaic susceptibility

Common and Scientific Names	*Increase of mites*	*Mosaic Susceptible*
A. Crop Plants		
Oats (*Avena sativa)*	none	yes
Barley (*Hordeum vulgare*)	poor	yes
Rye (*Secale cereale*)	poor	yes
Sorghum (*S. vulgare*)	fair-good	no
Sudangrass (*S. vulgare var. sudanense*)	poor	no
Corn (*Zea mays*)	poor-good	yes
Foxtail millet (*Setaria italica*)	poor	yes
Broom-corn millet or proso (*Panicum miliaceum*)	none	yes
Wheat (*Triticum aestivum*)	good	yes
B. Annual Grasses		
Jointed goatgrass (*Aegilops cylindrica*)	fair-good	yes
Wild Oats (*Avena fatua*)	none	yes
Japanese chess (*Bromus japonicus*)	none	yes
Cheat (*B.secalinus*)	good	yes
Downy chess (*B.tectorum*)	none	yes
Field sandbur (*Cenchrus pauciflorus*)	good	yes
Smooth crabgrass (*Digitaria ischaenum*)	fair-good	yes
Hairy crabgrass (*D. sanguinalis*)	none	yes
Barnyard grass (*Echinchloa crus-galli)*	poor	yes
Goosegrass (*Eleusine indica*)	none	no
Stinkgrass (*Eragrostis cilianensis*)	poor	yes
Teosinte (*Euchlaena mexicana*)	poor	no
Foxtail Barley (*Hordeum jubatum*)	poor	no data
Witchgrass (*Panicum capillare*)	none	yes
Yellow foxtail or bristlegrass (*Setaria lutescens*)	none	no
Bristly or bur foxtail (*S. verticillata*)	poor	yes
Green foxtail (*S. viridis*)	poor	yes
C. Perennial Grasses		
Tall wheatgrass (*Agropyron sp.*)	none	no
Western wheatgrass (*A. smithii*)	poor-fair	no
Slender wheatgrass (*A. trachycaulum*)	none	no data
Crested wheatgrass (*A.desertorum*)	none	no data
Meadow foxtail (*Alopecurus pratensis*)	none	no

Tall oatgrass (*Arrhenatherum elatius*)	poor	no
Blue grama (*Bouteloua gracilis*)	none	no
Side-oats grama (*B. curtipendula*)	none	no
Grama (*B. sp.*)	good	yes
Smooth brome (*Bromus inermis*)	very poor	no
Buffalograss (*Buchloe dactyloides*)	none	no data
Orchardgrass (*Dactylis glomerate*)	none	no
Canada wildrye (*Elymus canadensis*)	fair	yes
Indian ricegrass (*Orzyopsis hymenoides*)	poor-fair	yes
Switchgrass (*Panicum virgatum*)	none	no
Reed canarygrass (*Phalaris arundinacea*)	none	no
Canada bluegrass (*Poa compressa*)	poor	yes
Wheeler bluegrass (*P. nervosa*)	poor-fair	no
Bulbous bluegrass (*P. bulbosa*)	poor	yes
Bulbous bluegrass (*P. stenantha*)	poor	yes
Johnsongrass (*Sorghum halepense*)	good	no
Indian grass (*Sorghastrum nutans*)	none	no
Sand dropseed (*Sporobolus cryptandrus*)	none	no data
Green needlegrass (*Stipa viridula*)	none	no data
Needle-and-thread (*S. comata*)	poor-fair	no data
Pairie sandreed (*Calamovilfa longifolia*)	none	no data

developed from shattered grain. Volunteer wheat, therefore, becomes very important to sustaining mite populations. From these plants, which can become infected by mites feeding and reproducing, the viruliferous mites can be wind-borne into the wheat field. If weather conditions are warm, the increase of mites in the field can be great and the virus can be extensively inoculated. Fall infected wheat is the most seriously affected by the virus.

Control measutres

Control measures for wheat streak mosaic are aimed at destroying the populations of mites that transmit the WSMV and destroying the plants, which are the virus source. Taking the following measures best does this:

1. Destroy all volunteer cereals, old cereal stubble, and weed grasses in adjoining fields two weeks before planting, and three to four weeks before sowing in the field to be seeded. Doing this eliminates the mite vector as well as the mosaic-infected plants. The best control results when all wheat farmers in a community cooperate in destroying volunteer wheat and old stubble well ahead of planting time.

2. Sow winter wheat as late as practical after the Hessian fly-free date or the latest recommended date to escape migrations of the mite from corn, volunteer wheat or barley, or weed grasses. If winter wheat is not up until October or later, it usually escapes severe infestation, unless fall temperatures are above normal. An infection of winter cereals by wheat streak mosaic in the spring does relatively little damage.
3. Disease reactions may vary from one locality to another and from year to year, depending on the physiologic races of the pathogens present.
4. Chemical control of the wheat curl mite has not been successful. The tightly rolled and trapped leaves provide a natural protection for the mite, preventing contact with miticides. It is also difficult to know exactly when to apply chemicals for control.

Wheat Soil-borne mosaic (Symptoms)

Plants infected with Soil-borne Wheat Mosaic Virus (SBWMV) can show main two types of symptoms.

1. The first is leaf mottling, which appears as a light green and light yellow mosaic on the leaves. The mottling will be seen only very early in the season.
2. The second symptom is stunting to the point where the wheat plant looks like a rosette when growth begins in the spring. Under good growing conditions the infected plants may recover somewhat.
3. Fields may be uniformly diseased or more often may have spots in the field with virus symptoms.
4. Symptoms are most prominent in early spring. Warming spring temperatures tend to slow disease development and eventually the symptoms are almost completely masked.

The Virus

Soil-borne Wheat mosaic is caused by the soil-borne wheat mosaic virus (SBWMV) Particles (virions) of the soil-borne wheat mosaic virus (SBWMV) are hollow rigid rods 20 nm wide and of two lengths, 110 to 160 and 280 to 300 nm. The longer rods resemble the particles of tobacco mosaic virus. Particle lengths differ somewhat with strains of the virus, but both short and long rods appear to be necessary for infection.

The Vector

In nature, the way of transmission is via a soil-borne fungus, Polymyxa graminis Led. This fungus is a parasite in the roots of many higher plants, including wheat. The fungus enters roots through root hairs and epidermal cells of the roots under wet soil conditions by way of motile zoospores. The fungus carries the virus into the roots in which it colonizes. The virus is released into the plant tissues, proceeding in cell-to-cell "takeover."

Disease cycle

An unusual aspect of this disease is its mode of transmission to wheat plants. The virus is transmitted to the plant by a soil borne fungus P. graminis. The virus survives in the soil in close association with this fungus. Soils may remain infested with the fungus/virus for many years. When the fungus enters wheat roots, it transmits the virus. The fungus is a water mold and favors low, wet areas of the field, and this is usually where the disease is first seen. So the fungus infects wheat roots during cool, wet periods in the fall and possibly in the early spring. Fall infections permit the virus to elaborate to damaging proportions and predispose plants to other diseases and winter injury. SBWMV is not commonly a yield-reducing disease because higher spring temperatures inactivate the virus, and then symptoms do not appear on new leaves. Normally, spring infections occur too late to cause very much injury before warmer temperatures and crop maturation inhibit virus development. The virus and its fungus vector appear to be spread by cultivation, wind, water and other factors that cause movement of infested soil within a field. Yield reductions with SBWMV are uncommon except where extremely susceptible plants are present. Most wheat varieties are resistant to this pathogen, although that can vary.

Control measures

1. The most common and practical method of virus control is to use plant resistant wheat varieties. These varieties do not allow virus replication to occur, and the infection is stopped early.
2. Other control measures are directed at reducing the time the plants are in the field when vectors are active-thus the recommendation to plant after the fly-free date, when insect activity is reduced.
3. Systemic insecticide seed treatments have also shown some success.
4. Crop rotation and planting times may have minimal benefits.

Other virus or Virus-like diseases of wheat

Brome mosaic virus (BMV), which was described on bromegrass, is capable of affecting wheat, oats, corn, barley and rye. However, natural infections of wheat are of little economic importance. Some

Table 2.2. Identification of Wheat Viruses.

	Barley Yellow Dwarf	*Wheat Streak Mosaic*	*Soil-Borne Mosaic*	*Spindle Streak Mosaic*	*Maize Dwarf Mosaic*
When symptoms are observed	6-8 weeks after spring growth begins-yellowing at maturation, yellow to reddish flag leaves	4-6 weeks after spring growth begins. In fall on volunteer wheat.	Early spring. 1-2 weeks after spring growth begins. Rarely in fall.	Early spring. 1-2 weeks after spring growth begins.	Rare in wheat
Pattern in field	Random circular areas	Often along edgesof fields or near volunteer wheat. Diminishing with distance.	Somewhat circular areas, especially in wet areas.	More widespread in field than soil-borne mosaic.	Random plants
Leaf symptoms	Leaf tips bright yellow on upper leaves, or reddish flag leaves - most distinct.	Bright yellow leaves with streaking patterns toward tips. Most prominent on upper leaves.	Pale yellow leaves with mosaic patterns	Yellow-green mottling dashes and streaks parallel with veins	Mosaic, mottling

		Curling of upper leaves.			
Stunting	Some-but hard to identify; fewer tillers	Severe stunting from fall infections	Some stunting but also some recovery after warm weather	Mild stunting and fewer tillers	Mild stunting
Other symptoms	Poor roots; susceptible to winter injury	Poor roots; wilting; prostrate tillers	Poor roots; winter injury	Poor roots; winter injury	Incons-picuous
Vectors	Several aphid species	Wheat curl mite *Aceria tulipae*	Fungus *Polymyxa graminis*	Fungus *Polymyxa graminis*	Several aphid species
Conditions favoring infection	Long warm fall	Long warm fall. Early volunteer wheat from hail or other factors at harvest.	Wet soil in fall. Cool temperat-ures in spring.	Wet soil in fall. Cool temperat-ures in spring.	Proximity to infected Johnson-grass
Control	Resistant or tolerant varieties	Destroy volunteer wheat. Delay planting to "fly free date."	Soft red winter wheat varieties usually more tolerant than hard red winter wheat varieties	Resistant cultivars. Late planting.	Not needed

wheat cultivars are symptomless carriers of BMV, while others may develop streak-like mottle symptoms. Maize dwarf mosaic (MDM) is a serious disease of corn. The maize dwarf mosaic virus (MDMV) is a strain of the sugarcane mosaic virus and is, of course, most important on corn and sorghum. However, it can be transmitted to wheat by aphids. MDMV in wheat induces mild leaf mottling. It will be difficult to identify. Tobacco mosaic (TM), caused by the tobacco mosaic virus (TMV), is a prevalent and stable virus in nature with a wide host range, especially among dicotyledonous species. It can, however, infect certain grasses, among them wheat. In the case of a double infection when another wheat virus is involved, some additional symptom expression could be observed. Certain virus-like diseases caused by mycoplasma-like organisms (MLO's). Symptoms may be similar to other known viral diseases, or they may be inconspicuous. The ultra-microscopic organisms are more like bacteria than viruses. They do not have rigid cell walls, but have a fine outer membrane that allows greater flexibility than bacteria. Since Aster yellows has a wide host range and is very prevalent in nature, wheat can ostensibly have aster yellows infections. It is leaf hopper transmitted. Essentially, it is of minimal consequence to wheat production.

Viral Diseases of Pepper

Symptoms

1. They may also develop inward rolling and distorted tips. On fruits, solid circular spots may develop.
2. In many instances, they become rough, deformed and severe reduction in size.
3. When plants are affected when still young, they are stunted and express severe symptoms.

Disease cycle

The pepper mild mottle virus causes pepper mild mottle disease. The host range of the viruses consists of more than 750 plant species in various families including many vegetables such as tomato, cucumbers and legumes. The extensive host range serves as a source of infections and contributes to the spread of the viruses to the crop. The viruses are introduced to cultivated peppers mainly by aphid species after they have fed and acquired the viruses from wild reservoir hosts. More than sixty aphid species, including the common Aphis gossypi and Myzus persicae, are capable of transmitting viruses in a non-persistent manner.

Control measures

1. The pepper should not be planted near source plants such as Solanum nigrum and hosts such as tobacco, tomato and cucumber.
2. Besides using the recommended methods such as planting disease-free seedlings and uprooting the diseased plants, the fields and nurseries should be protected from the aphids.
3. Using sticky yellow polythene sheets, which are erected vertically on the windward side of the fields and nurseries, can help ward off these vectors. The aphids are attracted to the yellow colour and are caught on the sticky polythene.
4. The most effective control method is the use of virus-resistant varieties.

Tobacco Mosaic Virus Disease of Tomato and other Plants

The plant disease caused by tobacco mosaic virus is found worldwide. The virus is known to infect more than 150 types of herbaceous, dicotyledonous plants including many vegetables, flowers, and weeds. Infection by tobacco mosaic virus causes serious losses on several crops including tomatoes, peppers, and many ornamentals. Tobacco mosaic virus is one of the most common causes of virus diseases of plants in the world. Many viruses produce mosaic-like symptoms on plants.

Symptoms

1. Mosaic-like symptoms are characterized by intermingled patches of normal and light green or yellowish colors on the leaves of infected plants. Symptoms include various degrees of chlorosis, curling, mottling mosaic, dwarfing, distortion, and blistering of the leaves.
2. Tobacco mosaic damages the leaves, flowers, and fruit and causes stunting of the plant. Many times the entire plant is dwarfed and flowers are discolored.
3. The virus almost never kills plants but lowers the quality and quantity of the crop, particularly when the plants are infected while young.
4. Virus-infected plants often are confused with plants affected by herbicide or air pollution damage, mineral deficiencies, and other plant diseases.
5. Symptoms can be influenced by temperature, light conditions, nutritional factors, and water stress.

Common Plant Hosts

Common plant hosts for the mosaic virus are tomato, pepper, petunia, snapdragon, delphinium, and marigold. Tobacco mosaic virus also has been reported to a lesser extent in muskmelon, cucumber, squash, spinach, celosia, impatiens, ground cherry, phlox, zinnia, certain types of ivy, plantain, nightshade, and jimson weed. Although tobacco mosaic virus may infect many other types of plants, it generally is restricted to plants that are grown in seedbeds and transplanted or plants that are handled frequently.

Symptoms of tomato plants

1. In tomatoes, the foliage shows mosaic (mottled) areas with alternating yellowish and dark green areas.
2. Leaves are sometimes fern-like in appearance and sharply pointed.
3. Infections of young plants reduce fruit set and occasionally cause blemishes and distortions of the fruit.
4. The dark green areas of the mottle often appear thicker and somewhat elevated giving the leaves a blister-like appearance.

The Virus and Disease cycle

Viruses differ from fungi and bacteria in that they do not produce spores or other structures capable of penetrating plant parts. Since viruses have no active methods of entering plant cells, they must rely upon mechanically caused wounds, vegetative propagation of plants, grafting, seed, pollen, and being carried on the mouthparts of chewing insects. Tobacco mosaic virus is most commonly introduced into plants through small wounds caused by handling and by insects chewing on plant parts. The most common sources of virus inoculum for tobacco mosaic virus are the debris of infected plants that remains in the soil and certain infected tobacco products that contaminate workers hands.

Cigars, cigarettes, and pipe tobaccos can be infected with tobacco mosaic virus. Handling these smoking materials contaminates the hands, and subsequent handling of plants results in a transmission of the virus. Therefore, do not smoke while handling or transplanting plants. Once the virus enters the host, it begins to multiply by inducing host cells to form more viruses. Viruses do not cause disease by consuming or killing cells but rather by taking over the metabolic cell processes, resulting in abnormal cell functioning. Abnormal metabolic functions of infected cells are expressed as mosaic and other symptoms as previously described. Infected plants serve as reservoirs for the virus and the virus can be transmitted easily (either mechanically or by insects) to healthy plants.

Control measures

1. Unlike fungicidal chemicals used to control fungal diseases, to date there are no efficient chemical treatments that protect plant

parts from virus infection. Additionally, there are no known chemical treatments used under field conditions that eliminate viral infections from plant tissues once they do occur.

2. Practically speaking, plants infected by viruses remain so. Thus, control of tobacco mosaic virus is primarily focused on reducing and eliminating sources of the virus and limiting the spread by insects. Tobacco mosaic virus is the most persistent plant virus known. It has been known to survive up to 50 years in dried plant parts. Therefore, sanitation is the single most important practice in controlling tobacco mosaic virus.
3. The most common method of transferring the virus from plant to plant is on contaminated hands and tools. Workers who transplant seedlings should refrain from smoking during transplanting and wash their hands frequently and thoroughly with soap and water.
4. Tools used in transplanting can be placed in boiling water for 5 minutes and then washed with a strong soap or detergent solution. Dipping tools in household bleach is not effective for virus decontamination.
5. Any seedlings that appear to have mosaic symptoms or are stunted and distorted should be removed and destroyed. After removing diseased plants, never handle healthy plants without washing hands and decontaminating tools used to remove diseased plants.
6. Persons purchasing small tomato plants for transplanting should beware of any plants showing mottling, dwarfing, or stunting. Avoid the purchase of any affected plant.

Control method for Commercial Producers of tomato plants

Commercial greenhouse producers of tomatoes should follow control practices for seedling production as stated above. It is essential for commercial growers to constantly inspect and rogue diseased production plants while the plants are in the seedling stage. An experienced individual, who is familiar with the tobacco mosaic virus symptoms, should do the initial inspection. Roguing of young production plants is recommended and should take place before workers are allowed to prune or tie up production plants. When removing diseased plants, also remove one plant on either side of the diseased one. The reason for this is that it is almost impossible to remove a diseased plant and not contaminate the healthy adjacent plants. Never attempt to transplant a healthy tomato into the soil from which a diseased plant was removed. Roots from diseased plants will remain in the soil and provide the virus inoculum for the new transplant. As a matter of routine, soils from which production plants have been removed, following harvest should be steam sterilized before the introduction of new seedlings. Steam or air-steam mixtures can accomplish steam sterilization. In the preparation of soil for steam sterilization, sift it to remove clumps and large pieces of organic matter. The total soil mixture will have to be heated to a temperature of 200° F for 40 minutes. Since high temperatures are required, steam sterilization must be done in an enclosed system. High temperature thermometers to make sure the desired temperature has been reached should monitor temperatures within the steam sterilization system. Steam sterilization of soil also will eliminate fungi, insects, nematodes, and weeds from the soil. Steam sterilization also is recommended for gravel mixtures used in hydroponic operations following the same procedure described above. Grow individual production plants in separate containers so that the soil or growing media can be removed when roguing infected production plants. Remember that the soil harbors old root tissues that may serve as inoculum when new roots are introduced. Growing production plants in separate containers is also useful for the control of root diseases caused by fungi and bacteria.

Plum Pox Potyvirus (PPV) Disease of Stone Fruits

Symptoms

1. In Prunus, plum pox virus (PPV) symptoms appear on leaves, fruits, flowers, and seeds.
2. The severity of the symptoms varies according to the Prunus species and cultivar, PPV strain, season and location.
3. Leaves and fruit show chlorotic (yellowing) and necrotic (browning) ring patterns, and chlorotic bands or blotches.
5. Leaves and fruit also can be absent of symptoms, or have symptoms that are ameliorated during the growing season.
6. The fruit of apricot and plum can be misshapen and deformed or rings may be present on their stones.
7. Some peach cultivars may show color-breaking symptoms on the flower petals.
8. Virus infection can cause considerable losses. About 100 million stone fruit trees in Europe are currently infected, and susceptible cultivars can result in 80-100% yield losses. In eastern and central Europe, sensitive plum varieties can exhibit premature fruit drop and bark splitting.
9. Some sweet cherry fruits develop chlorotic and necrotic rings, notched marks, and premature fruit drop.

Table 2.3. Plum Pox Status and Associated Geographic Distribution

Disease Status	*Country*
Restricted Distribution	Albania, Austria, Cyprus, Czech Republic, France, Italy, Luxem-bourg, Moldova, Norway, Portugal, Southern Russia, Slovenia, Spain, Syria, Turkey, Ukraine, United Kingdom, United States
Widespread	Bulgaria, Croatia, Germany, Greece, Hungary, Poland, Romania, Slovakia, Former Yugoslavia
Introduced, Established	Azores, Bosnia-Herzegovina, Egypt, Former Russia, India, Lithuania
Introduced, Presumably Eradicated	Belgium, Netherlands, Switzerland
Present Status Unknown	Chile, Denmark

The Virus

Plum pox virus belongs to the genus Potyvirus in the family Potyviridae. Members of the genus Potyvirus have virions which are flexuous filaments with no envelopes, are aphid-transmitted in a non-persistent, stylet-borne manner, mechanically transmitted, and may or may not be seed-transmitted.

PPV has a single molecule of positive sense, ssRNA, that is 9.7 kb; virions are approximately 764 X 20 nm. The genome is expressed as a 350 kDa polyprotein precursor that is proteolytically processed by viral and host proteases into seven smaller functional proteins including a 3' coat protein and a helper component. It is the only recognized potyvirus infecting Prunus. The introduction of PPV to a new country or region is usually through propagative materials and the subsequent distribution of contaminated materials. The secondary spread can be rapid and results from aphid transmission.

Disease cycle

Plum pox virus has been transmitted by at least 20 aphid species, although only 4-6 are considered important vectors. The efficiency of transmission is dependent on the virus strain, host cultivars, age of the host cultivars, aphid species, and time of year. The most important aphid vectors reported from several countries are Brachycaudus cardui,

B. helichrysi, Myzus persicae, and Phorodon humuli. Reports vary from country to country, however, natural virus spread is low in July and August but high in spring and autumn. Spring flights of B. helichrysi, M. persicae, and P. humuli are most important for spread within and between orchards. Analysis of spacial distribution of PPV by Gottwald et al. (1995) suggest a lack of movement by aphid vectors to immediately adjacent trees and a preference for movement several tree spaces away. Aphids can acquire the virus in probes as short as 30 seconds, and can transmit for up to 1 hour. Aphids that have been starved before feeding can transmit for up to 3 hours after acquisition. There is no correlation between the ability to transmit PPV and the ability to colonize Prunus. PPV can be spread in orchards by transient aphids as efficiently as aphids colonizing Prunus (Labonne et al., 1995). Aphids were found to transmit PPV within 100-120 m of the source plants, but they have been shown to carry the virus on their stylets for several kilometers if starved during flight.

The plant host

Plum pox virus has a broad experimental host range although it has a rather restricted natural host range within the genus Prunus. It infects peaches, plums, apricots, nectarines, almonds, and sweet and tart cherry. Virus isolates vary in their reaction to different hosts, and not all strains or isolates infect the same host range. Prunus species that have been proven to be hosts in nature, or by inoculation trials followed by back transmissions include:

Prunus armeniaca - Apricot

P. persica - Peach

P. persica var. nectarina - Nectarine

P. domestica - Garden plum (prune)

P. salicina - Japanese plum

P. insititia - Damson plum

P. cerasifera - Myrobalan plum

P. glandulosa - Dwarf flowering almond, Cherry almond

P. avium - Sweet cherry

P. cerasus - Sour (tart) cherry

P. amygdalus - Almond

Wild Prunus may serve as an important secondary host of PPV and can have an impact on plum pox epidemiology and control (Polak, 1997). In addition to the above natural hosts, several wild Prunus species are susceptible:

P. spinosa - Blackthorn

P. americana - American plum

P. bessey - Western sand cherry

P. mahaleb - Mahaleb or St. Lucie cherry

P. mume - Japanese apricot

P. pumila - Sand cherry

P. hortulana - Hortulan plum

P. davidana - David peach, Chinese wild peach

P. tomentosa - Nanking cherry

P. nigra - Canada plum

P. maritime - Beach plum

P. laurocerasus - English cherry-laurel

Many non-Prunus species, in at least nine plant families, have been infected artificially with one or more strains of the plum pox virus, and in some cases found naturally infected in the field. Most of these are herbaceous annuals but a few are perennial or woody and could serve as over-wintering sources of the virus. Some of the common hosts include:

Campanula rapunculoides

Chenopodium quinoa

C. species

Lamium album

L. amplexicaule

L. purpureum

Lupinus albus

Lycium barbarum

L. halimifolium

Medicago lupulina

Melilotus officinalis

Ranunculus acer

R. arvensis

R. repens

Silene vulgaris

Solanum dulcamara

Trifolium incarnatum

T. pratense

T. repens

Zinnia elegans

Z. violacea

In addition the following are the more important herbaceous indicator or propagation hosts of plum pox virus:

Chenopodium foetidum

Nicotiana benthamina

N. bigelowii

N. clevelandii

N. occidentalis

N. edwardsonii

N. megalosiphon

N. tabacum

N. physalodes

Pisum sativum cv. Colmo

The virus is unevenly distributed in trees that are newly infected or have some degree of resistance, however, once an infection is established it can reach high titers in plant tissues such as leaves, flowers, and fruit in the spring and early summer. Several polyclonal and monoclonal antibodies have been developed and are used worldwide for PPV detection.

Control measures

There is no specific control measures are available at present to control this disease.

Viral Diseases of Potato

(a) Leaf roll of potato: (Symptoms)

1. The margins or the tips of the leaves become yellow.
2. The leaves of the infected plants roll up along the margins, leaving the midrib at the bottom of the trough.
3. In plants with diseased seed tubers, this rolling of leaves starts in the lower leaflets and progresses upward throughout the entire plant.
4. In some varities individual leaflets have a tendency to be more erect, giving the diseased plant an upright appearance.
5. As a result of infection, necrosis occurs in the phloem and there is excessive accumulation of starch.
6. The number of tubers produced per plant and their sizes in a diseased crop are greatly reduced.

Disease cycle

This disease caused by leaf roll virus also known as potato virus I or Solanum virus 14. Virus is not transmitted through the sap. Transmission of the virus in nature occurs through infected seed tubers and through the agency of insects. The main insect vector is an aphid Myzus persicae. The virus persists in the body of Myzus persicae for many days even if the vector has fed on many immune hosts after feeding on the diseased plants.

Host plants

Datura stramonium

Datura tulula

Physalis angulata

Physalis floridana

Lycopersicon esculentum

Control measures

1. Potato seed certification and roguing are recommended for disease control.
2. Some results of experiments suggest that if the tubers are stored at 37.5 C in a humid atmosphere for 20-25 days, they become free of the leaf roll virus.

(b) Mild mosaic of potato: (Symptoms)

1. Although the disease occurring widely in most of the seed stocks, causes very slight or negligible symptoms of interveinal mottling or mosaic.
2. Under favourable growth conditions of the host of the symptoms are masked, thus making recognition of the disease difficult.
3. The mosaic symptoms are almost completely masked at high temperatures (above 21^0C).
4, The virus is carried by the plants without any visible symptoms in many cells.
5. There is only slight dwarfing of the plant or deformation of the foliage.
6. The virus produces top necrosis in some verities.

Disease cycle

Mild mosaic of potato is caused by potato latent virus, potato mottle virus, potato virus X or Solanum virus I. The virus contains a large number of strains. The particles of the virus are long., flexuous rods (515 mu). The virus is easily transmitted by sap inoculation and is one of the few viruses, which spreads in the feild by the contact between healthy and diseased plants.

Virus can be transmitted by core grafting and by dodder, Cascuta campestris. No insect vector is known. Hower, the source of perennation and spread in the feild under natural conditions is mostly the diseased seed stock.

(c) Rugose Mosaic of Potato: (Symptoms)

1. The disease causes very severe damage to individual plants.
2. The foliage is not only mottled but is also severely wrinkled, puckered and markedly reduced in size.
3. The leaflet margins are rolled downwards and the entire plant is dwarfed.
4. The lower leaves generally have black necrotic veins.
5. While the symptoms of mottle may be masked by high temperature, roughness or rugosity and abnormal hairyness of leaves and dwarfing of the entire plant persist.
6. In severe cases plants may even die before producing any tuber.

Disease cycle

The disease ia caused by a combination of two viruses potato virus X (mild or latent mosaic) and potato virus Y. The latter causes blighting of the leaves in severe attacks and is also known as vein banding virus or Solanum virus 2. Under field condition of transmission of the virus is through small sized tubers. On diseased plants Myzus persiae is capable of efficiency transmitting virus Y which is one of the components of the complex.

Control measures

1. Systemic rouging of diseased plants as soon as they are noticed and their destruction by burring is recommended.
2. Indirect rouging of diseased seeds tubers is recommended at planting time by using large sized potatoes.

(d) Crinkle of Potato: (Symptoms)

1. The disease resembles rugose mosaic very closely but the yellowish patches on the foliage are bigger and hence, more prominent.
2. As the plant begins to die this colour becomes more pronounced and is accompanied by rusty spots, beginning near the tip of the leaves.
3. The foliage is brittle and is easily injured.

Viral Diseases of Papaya

(a) Leaf Curl of Papaya: (Symptoms)

1. Crinkling and curling of the leaves accompanied by vein clearing and reduction in the size of the leaves are the marked symptoms.
2. Leaves become leathery and brittle and the interveinal areas are raised on the upper surface due to hypertrophy, which produces rugosity.
3. The most prominent symptom is the downward and inward rolling of leaves and the veins get thickened and turn dark green in colour.
4. Sometimes all the leaves at the top of the plant areget affected by these symptoms
5. The petioles are twisted in a zigzag manner.
6. The severely affected plants fail to flower or bear fruits.
7. In the advanced stages of the diseases, defoliation takes place and plant growth is stunted.

Disease cycle

Tobacco virus 16 or Nicotiana virus causes this disease. Virus is not transmitted. It can be readily transmitted by grafting or juice inoculation. In nature the most important agent of its transmission is the white fly Bemisia tabaci.

Host plants

Papaya

Tomato

Tobacco

Zinnia

Althaea

Control measures

No suitable control method is known. Roguing is helpful in reducing the incidence of the disease.

(b) Mosaic of Papaya: (Symptom)

1. The disease is characterized by mosaic symptoms on the leaves with blister-like patches of green tissue distributed all over the yellowish green lamina.
2. Sometimes dark green spots and elongated streaks appearing like water-soaked areas, are formed on the petiol and the stem.
3. Fruits generally remain much smaller and get deformed. On some fruits large mosaic patches are also formed.

4. In severe infections, the plant has only a few small chlorotic or tendril-like leaves and they fail to flower.

Disease cycle

The papaya mosaic disease is caused by papaya mosaic virus. This virus is not seed borne. It can be transmitted by mechanical sap inoculation and wedge and bud grafting of aphids. Aphis malvae, Aphis gossypi and Aphis medicaginis are known as vectors. It is transmitted under natural conditions only through vectors.

Host plants

Trichosanthes anguina

Cucurbita maxima

Cucurbita pepo

Citrullus vulgaris

Cucumis sativus

Luffa acutangula

Lagenaria ciceraria

Viral Diseases of Cucurbits

Symptoms

1. Mosaics and stunting, reduced fruit yield.
2. Reduction of leaf laminae ("fernleaf").
3. Severe chlorosis, chlorotic local lesions, necrotic local lesions.

Host plants

Cucumis sativus

Lycopersicon esculentum

Spinacia oleracea

Chenopodium amaranticolor

Vigna unguiculata-.

Nicotiana edwardsonii

N. glutinosa

N. tabacum

Disease cycle

This disease is caused by cucumovirus. It is transmitted by a vector; an insect; more than 60 spp. including Acyrthosiphon pisum, Aphis craccivora and Myzus persicae; Aphididae. Transmitted in a non-persistent manner. Virus transmitted by mechanical inoculation; transmitted by seed (in 19 species, but in variable extents).

Synonyms of virus

Banana infectious chlorosis virus, coleus mosaic virus, cowpea banding mosaic virus, cowpea ringspot virus, cucumber virus 1, lily ringspot virus, pea top necrosis virus, peanut yellow mosaic virus, southern celery mosaic virus, soybean stunt virus, spinach blight virus, tomato fern leaf virus, pea western ringspot virus.

Viral Diseases of Okra

(a) Okra mosaic: (Symptoms)

1. Systemic symptoms appear on the youngest leaves as light green and regular veinal chlorosis.
2. Symptoms appear on two or three subsequent leaves but leaves produced later do not show any symptom.

Disease cycle

This disease is caused by Okra mosaic virus. This is reported from West Africa only. It belongs to tymoviruses group with ssRNA and is transmitted by sap inoculation but no vector is known.

(b) Yellow vein mosaic: (Symptoms)

1. The main symptom of this disease is vein clearing and veinal chlorosis of leaves.
2. The yellow network of veins is very conspicuous and the veins and the veinlets are thickened.
3. In severe cases the chlorosis may extend to the interveinal areas and may result in complete yellowing of the leaf.
4. Fruits are dwarfed, malformed and yellowing green in colour.

Disease cycle

The virus responsible for disease is Hibiscus virus Ior Yellow vein mosaic virus. It is transmitted by whitefly (*Bemisia tabaci*) and the virus is not sap transmissible. But under artificial conditions it could be transmitted by grafting.

Host plants

Croton sparsiflora

Malvastrum tricuspidatum

Ageratum sp.

Control measures

1. Protection of the crop from whiteflies and other insect vectors by spraying with Follidol (0.3 %) or other suitable insecticides reduces the incidence of the disease.

2. Destruction of weed hosts should also be given importance.
3. At present, no definite control measures are known. Variety of Sawani is reported to be tolerant.

Viral Diseases of Sugarcane

Edgerton (1968), Martin et al. (1961) and Hughes, et al. (1964) have described the following diseases of sugarcane as viral diseases:

(a) Mosaic
(b) Fiji
(c) Sereh
(d) Chorotic streak
(e) Streak
(f) Australian dwarf
(g) Dwarf
(h) Grassy Shoot
(i) Q.13 disease
(j) Ring mosaic
(k) Striate mosaic
(l) Sembur
(m) Spike

(a) Mosaic of Sugarcane: (Symptoms)

1. The first symptom of this disease appears about six weeks after planting and continue to develop throught the mansoon after which they become obliterated on maturity of the plant.
2. The primary and critical symptom of the disease is the appearance of pale patches or blotches in the green tissues of the leaf.
3. Small areas of the leaf are of a paler green colour than the rest. These patches are not uniform in size and shape.
4. The youngest unfolded leaves show the mottling very clearly while the symptoms are not very clear on older leaves.
5. Sometimes leaves of young tillers are found to be stiff, erect and crinkled.
6. Mottling of the stem also occurs in some varities and may lead to death of cells resulting in formation of cankers.

Disease cycle

The disease is caused by Sugarcane mosaic virus which belongs to potyviruses group. Its synonyms are Grass mosaic virus, *Saccharum* virus 1 and *Marmor sacchari*. The virus is transmitted from sugarcane to sugarcane by atleast seven species of aphids such as *Dactynotus ambrosiae*, *Hysteroneura setariae*, *Rhopalosiphum maidis*, *Toxoptera graminum*, *R. maidis*, and *Shizaphis graminum* are reported as vectors of sugarcane mosaic virus. Transmission is in the non-persistent manner.

Host plants

Maize

Sugarcane

Sorghum

Saccharum spontaneum

Control measures

1. Use only selected healthy setts for seed.
2. Eliminate the weed hosts.
3. Resistant varities are recommended for controlling the disease.

Other Viruses and their Diseases

Alfalfa Mosaic Virus (AMV)

Symptoms: Plants are mildly stunted and have whitish blotches on the leaves. The infected area is white-bleached and mottled. The fruit may be distorted.

Control: Aphids carry AMV. Peppers planted near alfalfa have a higher incidence of the disease. Control aphids and avoid planting near alfalfa. Use resistant varieties if they are available.

Beet Curly Top Virus (BCTV)

Symptoms: The most striking symptom is stunted plants and yellowing of the plants. The plants are also quite stiff and erect, and the leaves have a leathery feel.

Control: Curly top is carried by leafhoppers. Help prevent losses by partially shading plants with muslin tents or by other means early in the season. Leafhoppers that carry the virus do not feed in shady locations. Spraying or dusting with insecticide can be justified only when control is needed for other insects.

Cucumber Mosaic Virus (CMV)

Symptoms: Aphids carry CMV and cause stunted plants with dull green foliage with a leathery appearance.

Control: Control aphids and avoid planting near cucurbits.

Pepper Mottle Virus (PeMV)

Symptoms: Stunted plants, distorted fruit, and yield reduction are symptoms.

Control: Aphids carry PeMV. Control aphid vectors, practice good sanitation, and plant resistant varieties if available.

Potato Y (PVY)

Symptoms: Symptoms include mosaic and dark green vein-banding, leaf crinkle, leaf distortion, and plant stunting.

Control: PVY is carried by aphids. This has been called the most common pepper virus. Plant resistant varieties and control aphids.

Tobacco Etch Virus (TEV)

Symptoms: Mosaic and dark green vein-banding, leaf distortion, and plant stunting are symptoms. Tabasco plants wilt and die.

Control: Aphids carry TEV. Plant resistant varieties and control aphids.

Samsun Latent Tobacco Mosaic Virus (SLTMV)

Symptoms: Typical symptoms include mild mosaic and leaf distortion. Pods develop rings, line patterns, necrotic spots, and distortion. Plant stunting may also occur.

Control: Hands touching an infected plant and then touching an uninfected plant, so disinfecting hands with alcohol helps spread SLTMV mechanically. Clean seed and crop rotation also help control the virus.

Tomato Spotted Wilt Virus (TSWV)

Symptoms: TSWV is carried by thrips that feed on various virus-infected perennial flowering plants commonly grown around homes. Necrotic ringspots and leaf drop are symptoms.

Control: Use clean seed and control thrips.

Black Streaked Dwarf Virus (BSDV)

Symptoms: Stunting or dwarfing occurs, particularly when plants are infected in the seedling stage. Symptoms on pearl millet are not well-documented. On maize, white, waxy swellings occur on veins. Foliage is dark-green with chlorotic streaks and splitting of leaf margins.

Pathogen and disease characteristics: This belongs to Reoviridae fijivirus group. ISometric particles are 75-80 nm in diameter. Transmission occurs persistently by plant hoppers.

Host plants

Pearl millet

corn

rice

finger millet

barnyardgrass
Isachne globosa
barley
wheat
barnyard millet
Digitaria sanguinali
Geographic distribution: South Korea, Japan (on maize)

Alternative names for the disease

Rice black-streaked dwarf
Maize streaked dwarf
Seed transmission: Probably not seed transmitted.

Guinea Grass Mosaic Virus (GGMV)

Symptoms: Young diseased plants show lines of light green eye spots or a pale green mosaic, depending on cultivar. Symptoms develop into a striped mosaic by elongation and coalescing of the eye-spots. Some plants show severe symptoms with dwarfing.

Pathogen and disease characteristics: The Potyvirus is transmitted by aphids (*Hysteroneura setariae* and *Rhopalosiphum maidis*), probably non-persistently, but can be mechanically transmitted. Symptoms appear about 8 to 10 days after inoculation. Strain/host specificity may exist.

Host plants

Pearl millet
Bromus commutatus
Bromus macrostachys
Panicum crusgalli
Panicum maximum
Sorghum aroundinaceum
Zea mays
Paspalum conjugatum
Panicum maximum
Brachiaria brizantha
Brachiaria decumbens
Brachiaria dictyoneura
Brachiaria humidicola
Brachiaria jubata
Brachiaria ruzizensis

Brachiaria deflexa

Bromus arvensis

Bromus racemosus

Bromus sterilis

Coix lacryma-jobi

Echinochloa crus-galli

Oplismenus hirtelus

Panicum bulbosum

Panicum miliaceum

Paspalum racemosum

Setaria glaucum

Setaria italica

Setaria macrochaeta

Setaria verticillata

Stenotaphrum secondatum

Geographic distribution: Ivory Coast (from pearl millet)

Seed transmission: Not known to be seed transmitted.

Indian Peanut Clump Virus (IPCV)

Symptoms: Not available in the literature.

Pathogen and disease characteristics: The furovirus is vectored by soil borne fungus Polymyxa spp.

Host plants

Pearl millet

peanut

finger millet (Eleusine coracana)

foxtail millet (Setaria italica)

Geographic distribution: Natural distribution on pearl millet is not known. Occurrence has been confirmed in India.

Seed transmission: Very low rate of seed transmission (0.9%) has been observed in plants that had been grown in an infested field in India.

Panicum Mosaic Virus

Symptoms: Symptoms on pearl millet are expressed as a mild chlorotic mottle. On switchgrass, stunting can be severe in susceptible plants. Mild green mosaic and mottling, Yellow or light green blotchy mottling, mosaic, and streaking of leaves are characteristic. The entire plant or sectors of it can become chlorotic if badly stunted.

Pathogen and disease characteristics: 109S isometric virus, 28-30 nm in diameter. Single RNA (28S) and protein species (28,000 daltons). Six serotypes have been differentiated. A serological relationship exists between PMV and members of the phleum mottle virus group.

The virus is mechanically transmitted. PMV is a warm temperature virus. Incubation periods (7-18 days) are generally shorter at warmer, and longer at cooler temperatures. Optimum symptoms develop on many hosts with temperatures of 29 to 35^{0}C. Virus will remain infective in dessicated leaf tissue for up to 9 years.

Host plants

Switchgrass (Panicum virgatum L.),
broomcorn millet (P. miliaceum L.),
ticklegrass (P. capillare L.),
panicgrass (P. scribnerianum Nash),
Hall's panicum (P. hallii Vasey),
foxtail millet [Setaria italica (L.) Beauv],
barnyard grass [Echinochloa crusgalli (L.) Beauv.],
crabgrass [Digitaria sanguinalis (L.) Scop.]
Panicum ramosum,
P. decompositum,
P. turgidum,
Setaria verticillata
Setaria lutescens
Maize (Zea mays),
Panicum dichtomiflorum
Panicum virgatum L.
St. Augustinegrass
Stenotaphrum secundatum

Geographic distribution: Kansas. St. Augustine decline stain (PMV-SADV) occurs in Texas, Louisiana, Arkansas, South Carolina, and Mexico.

Synonyms: St. Augustine decline virus (SADV) is a strain of of PMV.

Seed transmission: Generally not known to be seed transmitted, however seed transmission of an SAD strain was reported in Setaria italica.

Satellite Panicum Mosaic Virus

Symptoms: Co-inoculation of Panicum Mosaic Virus (PMV) with its satellite virus (SPMV) results in a severe chlorotic mottle on pearl millet.

Pathogen and disease characteristics: The virus is a mechanically transmitted 42S isometric particle, 17 nm in diameter. Not infectious alone. Contains two RNA species (14 and 34S) and a single protein species (15,500 daltons). Serologically unrelated to panicum mosaic virus (PMV), but dependent upon PMV for replication. Two serotypes are known.

Host range: Host range is not well defined, but probably identical to that of panicum mosaic virus.

Geographic distribution: Kansas.

Seed transmission: Not known to be seed transmitted

Table 2.4. List of some Phytopathogenic virus.

S.No.	*Name*	*Taxon*
1.	Cucumber mosaic virus	Cucumovirus
2.	Turnip yellow mosaic virus	Tymovirus
3.	Brome mosaic virus	Bromovirus
4.	Potato virus X	Potexvirus
5.	Prunus necrotic ringspot virus	Ilarvirus
6.	Raspberry ringspot virus	Nepovirus
7.	Carnation mottle virus	Carmovirus
8.	Turnip mosaic virus	Potyvirus
9.	Lettuce mosaic virus	Potyvirus
10.	Cacao swollen shoot virus	Badnavirus
11.	Cacao yellow mosaic virus	Tymovirus
12.	Tobacco rattle virus	Tobravirus
13.	Beet yellows virus	Closterovirus
14.	Tobacco necrosis virus	Necrovirus
15.	Satellite virus	TNV satellite subgroup
16.	Arabis mosaic virus	Nepovirus
17.	Tobacco ringspot virus	Nepovirus
18.	Tomato ringspot virus	Nepovirus
19.	Prune dwarf virus	Ilarvirus
20.	Echtes Ackerbohnemosaik virus	Comovirus
21.	Carnation ringspot virus	Dianthovirus
22.	Red clover vein mosaic virus	Carlavirus
23.	Cocksfoot mottle virus	Sobemovirus
24.	Cauliflower mosaic virus	Caulimovirus
25.	Pea enation mosaic virus	Enamovirus
26.	Lettuce necrotic yellows virus	Cytorhabdovirus
27.	Cymbidium mosaic virus	Potexvirus

S.No.	Name	Taxon
28	Grapevine fanleaf virus	Nepovirus
29.	Broad bean stain virus	Comovirus
30.	Apple chlorotic leafspot virus	Trichovirus
31.	Apple stem grooving virus	Capillovirus
32.	Barley yellow dwarf virus	Luteoviridae
33.	Citrus tristeza virus	Closterovirus
34.	Wound tumor virus	Phytoreovirus
35.	Potato yellow dwarf virus	Nucleorhabdovirus
36.	Potato leafroll virus	Polerovirus
37.	Potato virus Y	Potyvirus
38.	Tomato black ring virus	Nepovirus
39.	Tomato spotted wilt virus	Tospovirus
40.	Bean yellow mosaic virus	Potyvirus
41.	White clover mosaic virus	Potexvirus
42	Tulare apple mosaic virus	Ilarvirus
43.	Squash mosaic virus	Comovirus
44.	Tobacco streak virus	Ilarvirus
45.	Narcissus mosaic virus	Potexvirus
46.	Alfalfa mosaic virus	Alfamovirus
47.	Cowpea mosaic virus	Comovirus
48.	Wheat streak mosaic virus	Tritimovirus
49.	Cowpea chlorotic mottle virus	Bromovirus
50.	Celery mosaic virus	Potyvirus
51.	Dahlia mosaic virus	Caulimovirus
52.	Belladonna mottle virus	Tymovirus
53.	Beet mosaic virus	Potyvirus
54.	Potato virus A	Potyvirus
55.	Tobacco etch virus	Potyvirus
56.	Papaya mosaic virus	Potexvirus
57.	Southern bean mosaic virus	Sobemovirus
58.	Cactus virus X	Potexvirus
59.	Cocksfoot streak virus	Potyvirus
60.	Potato virus S	Carlavirus
61.	Carnation latent virus	Carlavirus
62.	Sowthistle yellow vein virus	Nucleorhabdovirus
63.	Watermelon mosaic virus	Potyvirus
64.	Sowbane mosaic virus	Sobemovirus
65.	Robinia mosaic virus	Cucumovirus
66.	Potato spindle tuber 'virus'	Pospiviroid

S.No.	Name	Taxon
67.	Rice tungro virus	Waikavirus - Rice tungro spherical virus
68.	Barley stripe mosaic virus	Hordeivirus
69.	Tomato bushy stunt virus	Tombusvirus
70.	Plum pox virus	Potyvirus
71.	Tulip breaking virus	Potyvirus
72.	Maize rough dwarf virus	Fijivirus
73.	Bean common mosaic virus	Potyvirus
74.	Red clover mottle virus	Comovirus
75.	Poplar mosaic virus	Carlavirus
76.	Narcissus yellow stripe virus	Potyvirus
77.	Soil-borne wheat mosaic virus	Furovirus
78.	Carnation vein mottle virus	Potyvirus
79.	Tomato aspermy virus	Cucumovirus
80.	Cherry leaf roll virus	Nepovirus
81.	Broad bean wilt virus	Fabavirus
82.	Cucumber necrosis virus	Tombusvirus
83.	Apple mosaic virus	Ilarvirus
84.	Papaya ringspot virus	Potyvirus
85.	Broccoli necrotic yellows virus	Cytorhabdovirus
86.	Ryegrass mosaic virus	Rymovirus
87.	Potato virus M	Carlavirus
88.	Sugarcane mosaic virus	Potyvirus
89.	Beet western yellows virus	Polerovirus
90.	Cassava common mosaic virus	Potexvirus
91.	Parsnip mosaic virus	Potyvirus
92.	Peanut stunt virus	Cucumovirus
93.	Soybean mosaic virus	Potyvirus
94.	Maize mosaic virus	Nucleorhabdovirus
95.	Henbane mosaic virus	Potyvirus
96.	Lily symptomless virus	Carlavirus
97.	Pokeweed mosaic virus	Potyvirus
98.	Potato aucuba mosaic virus	Potexvirus
99.	American wheat striate mosaic virus	Cytorhabdovirus
100.	Rice transitory yellowing	Rhabdoviridae
101.	Broad bean mottle virus	Bromovirus
102.	Rice dwarf virus	Phytoreovirus
103.	Grapevine chrome mosaic virus	Nepovirus
104.	Pepper veinal mottle virus	Potyvirus

S.No.	*Name*	*Taxon*
105.	Wild cucumber mosaic virus	Tymovirus
106.	Black raspberry latent virus	Ilarvirus
107.	Cocksfoot mild mosaic virus	Sobemovirus
108.	Bean pod mottle virus	Comovirus
109.	Turnip crinkle virus	Carmovirus
110.	Chrysanthemum virus B	Carlavirus
111.	Clover yellow mosaic virus	Potexvirus
112.	Pea streak virus	Carlavirus
113.	Scrophularia mottle virus	Tymovirus
114.	Hydrangea ringspot virus	Potexvirus
115.	Eggplant mottled dwarf virus	Nucleorhabdovirus
116.	Iris mild mosaic virus	Potyvirus
117.	Hippeastrum mosaic virus	Potyvirus
118.	Agropyron mosaic virus	Rymovirus
119.	Sugarcane Fiji disease virus	Fijivirus
120.	Pea early-browning virus	Tobravirus
121.	Radish mosaic virus	Comovirus
122.	Passionfruit woodiness virus	Potyvirus
123.	Oat blue dwarf virus	Marafivirus
124.	Eggplant mosaic virus	Tymovirus
125.	Turnip rosette virus	Sobemovirus
126.	Strawberry latent ringspot virus	Nepovirus
127.	Elderberry latent virus	Carmovirus
128.	Okra mosaic virus	Tymovirus
129.	Parsnip yellow fleck virus	Sequivirus
130.	Pelargonium flower break virus	Carmovirus
131.	Clover yellow vein virus	Potyvirus
132.	Chicory yellow mottle virus	Nepovirus
133.	Maize streak virus	Mastrevirus
134.	Cowpea aphid-borne mosaic virus	Potyvirus
135.	Rice black streaked dwarf virus	Fijivirus
136.	Carnation necrotic fleck virus	Closterovirus
137.	Carrot mottle virus	Umbravirus
138.	Potato mop-top virus	Pomovirus
139.	Elm mottle virus	Ilarvirus
140.	Cowpea mild mottle virus	Carlavirus
141.	Peanut mottle virus	Potyvirus
142.	Mulberry ringspot virus	Nepovirus

S.No.	*Name*	*Taxon*
143.	Barley yellow mosaic virus	Bymovirus
144.	Beet necrotic yellow vein virus	Benyvirus
145.	Oat mosaic virus	Bymovirus
146.	Pea seed-borne mosaic virus	Potyvirus
147.	Bearded iris mosaic virus	Potyvirus
148.	Saguaro cactus virus	Carmovirus
149.	Rice yellow mottle virus	Sobemovirus
150.	Peach rosette mosaic virus	Nepovirus
151.	Tobacco mosaic virus (type strain)	Tobamovirus
152.	Ribgrass mosaic virus	Tobamovirus
153.	Sunn-hemp mosaic virus	Tobamovirus
154.	Cucumber green mottle mosaic virus	Tobamovirus
155.	Odontoglossum ringspot virus	Tobamovirus
156.	Tomato mosaic virus	Tobamovirus
157.	Wheat yellow leaf virus	Closterovirus
158.	Onion yellow dwarf virus	Potyvirus
159.	Cherry rasp leaf virus	Nepovirus
160.	Myrobalan latent ringspot virus	Nepovirus
161.	Bidens mottle virus	Potyvirus
162.	Sweet potato mild mottle virus	Ipomovirus
163.	Strawberry crinkle virus	Cytorhabdovirus
164.	Citrus leaf rugose virus	Ilarvirus
165.	Raspberry bushy dwarf virus	Idaeovirus
166.	Narcissus tip necrosis virus	Carmovirus
167.	Wheat spindle streak mosaic virus	Bymovirus
168.	Desmodium yellow mottle virus	Tymovirus
169.	Oat necrotic mottle virus	Rymovirus
170.	Narcissus latent virus	Macluravirus
171.	Clitoria yellow vein virus	Tymovirus
172.	Rice necrosis mosaic virus	Bymovirus
173.	Cacao necrosis virus	Nepovirus
174.	Raspberry vein chlorosis virus	Rhabdoviridae
175.	Pangola stunt virus	Fijivirus
176.	Artichoke Italian latent virus	Nepovirus
177.	Panicum mosaic virus	Sobemovirus
178.	Cymbidium ringspot virus	Tombusvirus
179.	Soybean dwarf virus	Luteoviridae
180.	Brome mosaic virus	Bromovirus

S.No.	*Name*	*Taxon*
181.	Red clover necrotic mosaic virus	Dianthovirus
182.	Carnation etched ring virus	Caulimovirus
183.	Orchid fleck virus	Rhabdoviridae
184.	Tobamovirus group	Tobamovirus genus
185.	Nepovirus group	Nepovirus genus
186.	Grapevine Bulgarian latent virus	Nepovirus
187.	Potato virus T	Trichovirus
188.	Rubus yellow net virus	Unassigned virus
189.	Nicotiana velutina mosaic virus	Unassigned virus
190.	Guinea grass mosaic virus	Potyvirus
191.	Dasheen mosaic virus	Potyvirus
192.	Bean golden mosaic virus	Begomovirus
193.	Kennedya yellow mosaic virus	Tymovirus
194.	Maize chlorotic dwarf virus	Waikavirus
195.	Daphne virus X	Potexvirus
196.	Frangipani mosaic virus	Tobamovirus
197.	Cowpea mosaic virus	Comovirus
198.	Raspberry ringspot virus	Nepovirus
199.	Comovirus group	Comovirus genus
200.	Potexvirus group	Potexvirus genus
201.	Lilac ring mottle virus	Ilarvirus
202.	Lilac chlorotic leafspot virus	Capillovirus
203.	Andean potato mottle virus	Comovirus
204.	Blueberry shoestring virus	Sobemovirus
205.	Sonchus yellow net virus	Nucleorhabdovirus
206.	Potato black ringspot virus	Nepovirus
207.	Beet yellow stunt virus	Closterovirus
208.	Satsuma dwarf virus	Nepovirus
209.	Cowpea severe mosaic virus	Comovirus
210.	Beet curly top virus	Curtovirus
211.	Alfalfa latent virus	Carlavirus
212.	Cowpea mottle virus	Carmovirus
213.	Cucumber mosaic virus	Cucumovirus
214.	Tymovirus group	Tymovirus genus
215.	Bromovirus group	Bromovirus genus
216.	Arracacha virus A	Nepovirus
217.	Oat sterile dwarf virus	Fijivirus
218.	Carrot thin leaf virus	Potyvirus
219.	Strawberry vein banding virus	Caulimovirus

S.No.	*Name*	*Taxon*
220.	Maize rayado fino virus	Marafivirus
221.	Chloris striate mosaic virus	Mastrevirus
222.	Erysimum latent virus	Tymovirus
223.	Broad bean necrosis virus	Pomovirus
224.	Lucerne transient streak virus	Sobemovirus
225.	Lucerne Australian latent virus	Nepovirus
226.	Citrus exocortis viroid	Pospiviroid
227.	Hibiscus chlorotic ringspot virus	Carmovirus
228.	Heracleum latent virus	Trichovirus
229.	Alfalfa mosaic virus	Alfamovirus
230.	Turnip yellow mosaic virus	Tymovirus
231.	Bean mild mosaic virus	Carmovirus
232.	Tobacco leaf curl virus	Begomovirus
233.	Hibiscus latent ringspot virus	Nepovirus
234.	Tobacco necrotic dwarf virus	Luteoviridae
235.	Peanut clump virus	Pecluvirus
236.	Melandrium yellow fleck virus	Bromovirus
237.	Blackgram mottle virus	Carmovirus
238.	Quail pea mosaic virus	Comovirus
239	Maclura mosaic virus	Macluravirus
240.	Leek yellow stripe virus	Potyvirus
241.	Hop mosaic virus	Carlavirus
242.	Potato virus Y	Potyvirus
243.	Cauliflower mosaic virus	Caulimovirus
244.	Plant rhabdovirus group	Rhabdoviridae family
245.	Potyvirus group	Potyvirus genus
246.	Bean rugose mosaic virus	Comovirus
247.	Viola mottle virus	Potexvirus
248.	Rice ragged stunt virus	Oryzavirus
249.	Carrot red leaf virus	Luteoviridae
250.	Shallot latent virus	Carlavirus
251.	Cereal chlorotic mottle virus	Rhabdoviridae
252.	Galinsoga mosaic virus	Carmovirus
253.	Pepper mottle virus	Potyvirus
254.	Avocado sun-blotch viroid	Avsunviroid
255.	Peru tomato virus	Unassigned virus
256.	Tephrosia symptomless virus	Carmovirus
257.	Pea enation mosaic virus	Enamovirus
258.	Tobacco etch virus	Potyvirus

S.No.	*Name*	*Taxon*
259.	Carlavirus group	Carlavirus genus
260.	Closterovirus group	Closterovirus genus
261.	Hop latent virus	Carlavirus
262.	American hop latent virus	Carlavirus
263.	Elderberry carlavirus	Carlavirus
264.	Foxtail mosaic virus	Potexvirus
265.	Helenium virus S	Carlavirus
266.	Plantain virus X	Potexvirus
267.	Blueberry leaf mottle virus	Nepovirus
268.	Beet leaf curl virus	Rhabdoviridae
269.	Rice stripe virus	Tenuivirus
270.	Arracacha virus B	Nepovirus
271.	Artichoke yellow ringspot virus	Nepovirus
272.	Pelargonium zonate spot virus	Unassigned virus
273.	Hypochoeris mosaic virus	Furovirus
274.	Southern bean mosaic virus	Sobemovirus
275.	Ilarvirus group	Ilarvirus genus
276.	Tulip virus X	Potexvirus
277.	Ullucus virus C	Comovirus
278.	Tobacco yellow dwarf virus	Mastrevirus
279.	Voandzeia necrotic mosaic virus	Tymovirus
280.	American plum line pattern virus	Ilarvirus
281.	Spinach latent virus	Ilarvirus
282.	Zucchini yellow mosaic virus	Potyvirus
283.	Maize white line mosaic virus	Unassigned virus
284.	Maize chlorotic mottle virus	Machlomovirus
285.	Artichoke vein banding virus	Nepovirus
286.	Bean leaf roll virus	Luteoviridae
287.	Coconut cadang-cadang viroid	Cocadviroid
288.	Asparagus virus 2	Ilarvirus
289.	Honeysuckle latent virus	Carlavirus
290.	Tomato ringspot virus	Nepovirus
291.	Potato leafroll virus	Polerovirus
292.	Papaya ringspot virus	Potyvirus
293.	Watermelon mosaic virus 2	Potyvirus
294.	Plant reovirus group	Reoviridae family
295.	Caulimovirus group	Caulimovirus genus
296.	Rice gall dwarf virus	Phytoreovirus
297.	African cassava mosaic virus	Begomovirus

S.No.	Name	Taxon
298.	Radish yellow edge virus	Alphacryptovirus
299.	Rice hoja blanca virus	Tenuivirus
300.	Maize stripe virus	Tenuivirus
301.	Olive latent ringspot virus	Nepovirus
302.	Melon necrotic spot virus	Carmovirus
303.	Tomato golden mosaic virus	Begomovirus
304.	Wineberry latent virus	Potexvirus
305.	Blackeye cowpea mosaic virus	Potyvirus
306.	Cherry leaf roll virus	Nepovirus
307.	Tobacco streak virus	Ilarvirus
308.	Carnation ringspot virus	Dianthovirus
309.	Tobacco ringspot virus	Nepovirus
310.	Iris fulva mosaic virus	Potyvirus
311.	Poinsettia mosaic virus	Tymovirus
312.	Barley yellow striate mosaic virus	Cytorhabdovirus
313.	Tobacco stunt virus	Unassigned virus
314.	Yam mosaic virus	Potyvirus
315.	Carnation cryptic virus	Alphacryptovirus
316.	Potato virus V	Potyvirus
317.	Velvet tobacco mottle virus	Sobemovirus
318.	Solanum nodiflorum mottle virus	Sobemovirus
319.	Cucumber leaf spot virus	Carmovirus
320.	Rice grassy stunt virus	Tenuivirus
321.	Sweet clover necrotic mosaic virus	Dianthovirus
322.	Northern cereal mosaic virus	Cytorhabdovirus
323.	Mung bean yellow mosaic virus	Begomovirus
324.	Iris mild mosaic virus	Potyvirus
325.	Tobacco vein mottling virus	Potyvirus
326.	Hop stunt viroid	Hostuviroid
327.	Blueberry red ringspot virus	Caulimovirus
328.	Ginger chlorotic fleck virus	Sobemovirus
329.	Subterranean clover mottle virus	Sobemovirus
330.	Pepper mild mottle virus	Tobamovirus
331.	Soybean chlorotic mottle virus	Caulimovirus
332.	White clover cryptic virus 2	Betacryptovirus
333.	Black raspberry necrosis virus	Unassigned virus
334.	Cassia yellow blotch virus	Bromovirus
335.	Dioscorea latent virus	Potexvirus
336.	Nerine virus X	Potexvirus

S.No.	*Name*	*Taxon*
337.	Bean common mosaic virus	Potyvirus
338.	Iris severe mosaic virus	Potyvirus
339.	Luteovirus group	Luteoviridae family
340.	Johnsongrass mosaic virus	Potyvirus
341.	Maize dwarf mosaic virus	Potyvirus
342.	Sugarcane mosaic virus	Potyvirus
343.	Lettuce necrotic yellows virus	Cytorhabdovirus
344.	Barley stripe mosaic virus	Hordeivirus
345.	Groundnut rosette assistor virus	Luteoviridae
346.	Tobacco rattle virus	Tobravirus
347.	Pepper ringspot virus	Tobravirus
348.	Miscanthus streak virus	Mastrevirus
349.	Apple scar skin viroid	Apscaviroid
350.	Pepino mosaic virus	Potexvirus
351.	Tobacco mild green mosaic virus	Tobamovirus
352.	Tombusvirus group	Tombusvirus genus
353.	Citrus tristeza virus	Closterovirus
354.	Potato virus X	Potexvirus
355.	Groundnut rosette virus	Umbravirus
356.	Barley mild mosaic virus	Bymovirus
357.	Cassava American latent virus	Nepovirus
358.	Squash leaf curl virus	Begomovirus
359.	Sorghum mosaic virus	Potyvirus
360.	Raspberry bushy dwarf virus	Idaeovirus
361.	Cassava Ivorian bacilliform virus	Unassigned
362.	Peach latent mosaic viroid	Avsunviroid
363.	Tospovirus genus	Tospovirus genus
364.	Peach rosette mosaic virus	Nepovirus
365.	Pear blister canker viroid	Apscaviroid
366	Potyviridae family	Potyviridae family
367	Narcissus late season yellow virus	Potyvirus

3

Diagnostic Features

Every plant, in its particular way, shows when it is suffering from disease. Those concerned with the growing of plants should become familiar with those outward signs of symptoms which exist when plants are sick. In the nature of the resulting symptoms, there are no sharp and fast lines that can be drawn between parasitic and non-parasitic diseases, or between various "insect" troubles and those caused by bacteria or fungi. The first thing of importance to the grower is to detect the presence of disease and then to learn its cause, the probable outcome and the need for or possibilities of control. The following brief survey of symptoms will be of service to the grower.

Discolouration or Change of Colour from the Normal

The most frequent discolourations are those affecting parts which are normally green, the presence of disease inducing a paling of the colour which may finally reach a complete disappearance of the green pigment. Instead of showing the green colour, the affected parts may be of a yellow cast, *etiolated* when the change is caused by insufficient light, or *chlorotic* when induced by other factors, while a complete absence of pigment results in *albinism*. A silvery sheen of normally green parts is also a characteristic accompaniment of some diseased conditions. Localized spots or areas showing whitish a gray, red a purple, brown, black, a variegated and concentrically zonate discolourations are signals of disease.

The above manifestations of disease should not be confused with the common red colouration of early spring foliage or the autumnal changes preceding leaf fall.

Shot Hole or Perforation of Leaves

The formation of localised lesions on leaves is frequently followed by the falling out of the dead tissue, leaving circular or irregular perforations which have suggested the term "shot hole." Certain species or varieties, especially of stone fruits, such as cherry, apricot or peach, are prone to develop the shot-hole symptom. Shot-hole development may result from bacterial or fungous pathogens or from the operation of non-parasitic factors such as frost injury, localised action of toxic chemicals or nutritional disturbances.

Wilting

The wilting of growing plants during a hot summer day followed by recovery is a normal occurrence and should not be confused with wilting as a symptom of disease. Two types of wilting from disease may be recognised: (*a*) the sudden falling over or "dropping dead" of young seedlings generally referred to as *damping-off*; and (*b*) the wilting without recovery of growing or adult plants. The juvenile wilting is caused by the invasion of a parasite in the young stem at or near the ground level, while the mature plants the water-conducting system ceases to function in the so-called "wilt" diseases.

Necrosis or Death of Parts

Localised groups of cells in certain organs may die and assume a brown colour as may be illustrated by the bitter pit or Baldwin spot of apples, or the internal brown spot of potato tubers. In other cases special organs of the plant such as leaves, stems or twigs, buds, flowers and developing fruits assume a dark or brown colouration characteristic of dead plant tissue, producing a condition frequently referred to as "blight" of the affected organ.

Dwarfing or Atrophy

Parts or organs such as leaves, fruits or flowers may be greatly reduced in size, suggesting such common names as "little leaf" disease of the apple or "little peach" with fruits or reduced size. In other cases, the entire individual is reduced in the size as a result of unfavourable factors, either environmental or parasitic.

Increase in Size or Hypertrophy

Practically any plant organs, such as roots, leaves, fruits, etc., may be stimulated by the action of a parasite or by other factors resulting in enlarged and sometimes malformed structures. This hypertrophy may result from an increase in size of the component cells or from an increased cell division forming a larger number of component cell (hyperlasia).

Transformation of Organs or the Replacement of Organs by New Structures

In the ergot of rye and other grasses (*Claviceps* spp.), certain of the ovaries are destroyed by the action of the parasite, and, in the place of the seeds, horny, elongated dark-purple spurlike bodies, the sclerotia or "ergots," appear. Stamens may become leafy in the greenear or downy mildew disease (*Sclerospora graminicola*) of *Pennisetum typhoideum* and other grasses; the whole staminate head, the tassel, or the ear, may become leafy in the head smut (*Sorosporium reilianum*) of corn; petals may become like sepals, stamens like carpels and carpels leaflike in the white rust (*Albugo candida*) of crucifers. The term "phyllody" is applied to the change of floral organs into leafy structures.

Mummification

The transformation of fruits into shriveled structures called "mummies" is a phase of numerous diseases affecting our commercial fruits and may also occur in the fruits of wild plants. The fruit generally rots first, and during the change becomes filled with the mycelium of the parasite, after which it shrivels and becomes somewhat dry. Mummies may remain on the tree or fall to the ground, and, sooner or later, they give rise to a crop of spores which serve to spread the parasite. The formation of mummies is a very characteristic feature of the brown rot (*Sclerotinia* spp.) of stone or pome fruits. Other cases of mummification may be found in bitter rot and black rot of the apple and in the black rot of the grape.

Alteration in Habit and Symmetry

Some plants which under normal conditions are more or less prostrate or creeping become ascending or even erect when attacked by a fungous parasite and, as it were, signal their distress. This is notably true in purslane (pusley), a common garden weed, when attacked by white rust and in certain spurges when harboring the cluster-cup stages of true rusts. In these and other cases a dorsiventral symmetry is changed to a more or less evident radial symmetry. The vegetative organs may shown various alternations; leaves may be changed from sImple to irregularly lobed, from the ovate from the ovate form to greatly elongated structures, or they may be variously twisted or deformed; stems may also be twisted or deformed, internodes elongated or shortened; and branching may be reduced or increased. Floral transformations are illustrated by the change of the club types of what heads to elongated ones by bunt or stinking smut, the flower head of an Acacia to a spike by a rust infection, regular flowers to irregular, cyclic to strobilate types, dioecious to perfect and various other modifications.

Destruction of Organs

The complete destruction of organs may result either from non-parasitic causes or from the inroads of a parasite. Oxules may dry up without the formation of seeds from non-parasitic causes, or seed may be destroyed by the action of a parasite as in smut of sheep sorrel or in the "bladder plums" due to *Taphrina pruni*. The so-called "seed" or fruits of cereals (the caryopses) are temporarily destroyed by the operation of the parasites of the various Kernel smuts (wheat, oats, sorghum), while in other cases the destruction is more complete and the entire inflorescence may be involved as in loose smuts (wheat. oats, barley).

Dropping of Leaves, Blossoms, Fruits or Twigs

This is, of course, to be considered as a symptom of disease only when it occurs prematurely or in excessive amount. Owing to a sudden change from a moist greenhouse to a dry room, from moderate to intense light or from cool conditions to warm, such house plants as fuchsia, foliage begonias, azaleas, rubber plants and many others may drop their leaves. The shedding of leaves from the action of parasites may be noted in the leaf casts of pine and larch, the leaf spot of alfalfa, and the leaf spot or yellows (*Coccomyces hiemalis*) of cherry. The shedding of blossoms may be illustrated bythe non-parasitic blossom drop of the tomato or by the so-called "shelling" of grape blossoms or of partially developed berries. Non-parasitic dropping of fruits may be illustrated by the shedding of bolls in cotton or the June drop of stone or pome fruits, and shedding of fruits from the action of a parasite by severe infections of scab in apples or California blight in cherries.

The Production of Excrescences and Malformations

The abnormal formations below may be included under this heading:

1. An abnormal development of hairs or trichomes.from the surfaces of leaves giving feltlike patches, first believed to be of fungous origin and named *erineum*, but now known to be caused by parasitic mites and designated *erinose*. Erinose spots on grape are white at first, but dirty brown which age, those on the mountain maple bright red or scarlet.

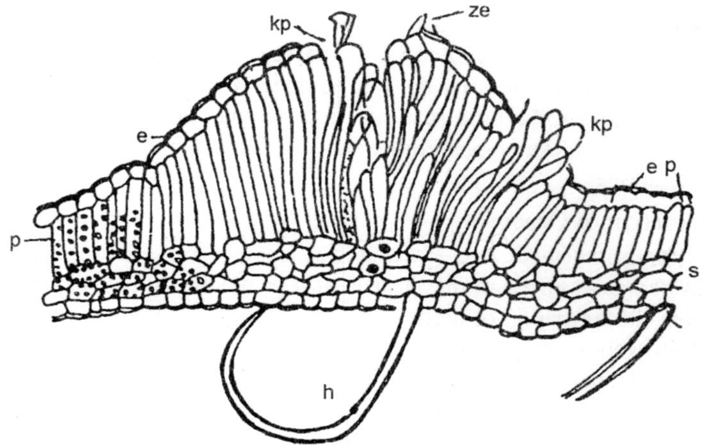

Fig. 3.1. Leaf intumescence of Cassia tomentosa.

2. *Intumescences* or pustulelike destentions of tissue, occurring most abundantly on leaves but also on stems or fruits, due to the abnormal elongation of groups of cells. If the overdevelopment of cells is somewhat general, rather than distinctly localized, and extensive swellings of organs result, the condition is spoken of as "edema" or dropsy. Both of these abnormalities are of non-parasitic origin.
3. *Galls*, or localised enlargements on various organs in the form of small pustules or warts, larger tubercles, tumors or masses of cells making a morbid outgrowth of either fleshy or woody character, in which host tissues, and parasites mingle. Some familiar illustrations in this group and the small or large tumors of corn smut (*Ustilago zeae*), the so-called "cular apples" (*Gymnosporangium* spp.) of our common red cedar, the elongated galls or "black knot" (*Dibotryon morbosum*) on twigs and branches of cherry and "crown gall," the woody or fleshy outgrowths from the crown and other portions of fruit trees and numerous other herbaceous or woody plants.
4. *Cankers*, or localized lesions on woody or, more rarely, herbaceous stems, which generally result in the corrosion and sloughing away of tissue with the final production of an open wound, exposing or penetrating the wood. Cankers may be annual or perennial, that is, increasing in size from season to season, and of either parasitic or non-parasitic origin. Typical Cankers are phases of the Pacific Coast and perennial cankers of apple; the chestnut-tree blight or Endothia canker of American chestnut; the European canker of

apple and the fire blight of apple, pear and other hosts, all of parasitic origin; while non-parasitic cankers are well-illustrated by "winter-sunscald" lesions which appear on the south-west side of tree trunks or on the exposed faces of larger limbs.

5. *Witches'-brooms* or "hexenbesen," closely grouped clusters of fine slender branches, generally arranged more or less parallel to each other, and frequently originating from an enlarged axis. These may be illustrated by the witches'-broom (*Taphrina cerasi*) of the cherry, rust brooms on various coniferous trees and brooms on pine, larch and fir caused by scaly or dwarf mistletoes (*Razoumofskya* spp.).
6. *Hairy root*, or an abnormal number of fine fibrous roots frequently making compact clusters. According to the location on the root or crown: *broom-root* type from a lateral root; *wooly knot* form with the roots from a crown-gall-like enlargement; an aerial with the roots or their primordia on the trunk or branches.
7. *Rosette*, or closely grouped clusters of leaves caused by the failure of plants to make a normal elongation. This should not be confused with the normal rosette habit of certain plants. This habit characterizes contain diseases as the apple rosette of the Pacific Northwest, the pecan- rosette of the southern United States and the rosette disease of peaches and plums in the southeastern states.
8. The development of dormant or rudimentary structures, or of entirely new organs or structures, either similar to or entirely unlike any normal parts of the host. Dormant buds may be started into growth, stamens normally rudimentary may grow to full size, extra petals may appear, or an entirely dissimilar outgrowth may be formed as an outgrowth from the frond of a fern (*Taphrina laurencia* of Pteris).
9. *Proliferation* or *prolification*, that is, the formation of a new growth from an organ after it has reached the form or stage which would normally end its development. Some typical illustrations are the formation of a needle-bearing shoot from the apex of a cone (larch); so-called "sprouting pears" in which one pear forms directly above another or a leaf-bearing twig grows out from the calyx end; or a new rose from the center of an old one.
10. *Rolling*, *crinkling* or *curling* of leaves may result from either parasitic or non-parasitic factors. Leaf curl (*Taphrina deformans*), a fungous disease of peach; leaf roll, a virous disease of potato; and frost blistering and curling of apple leaves are typical illustrations.

11. *Fasciation* and *spiralism* are not uncommon types of malformation, the former resulting in a change of a cylindrical organs to a broad, flattened and more or less bandlike structure, the latter characterised by the change of a straight, cylindrical. stem to form an irregular, scraggly structure with a flat or open spiral.
12. The abnormal roughening of surfaces, which under normal conditions should remain smooth. This may be illustrated by the russeting of fruits by frost or by sprays and by "scurf" or "scab" lesions caused by fungous parasites.
13. Fruits may be variously deformed, or root crops may be cracked, corroded or irregular in. shape. Such changes may result from either parasitic or nonparasitic disturbances. Apples may be greatly reduced in size and malformed by the scab fungus or by freezing when the fruits were very young, but it should be kept in mind that insect pests also may be the cause of deformities.

Production of an Exudate

The production of an exudate from certain plant organs is an accompaniment of certain diseased conditions, but should not be confused with the normal physiological processes of *guttation* or the forcing out of cell sap on a free surface, or with bleeding from cut ends of stems or other plant organs. Exudates may be classified according to their causes and their composition as:

(a) *Bacterial* exudates, well-illustrated by ooze from lesions, of the fire blight on apple, pear an other hosts.

(b) *Slime flux*, a semifluid or fluid outflow from the bark or wood of various deciduous trees.

(c) *Gummosis*, resulting in formation of clear or amber-coloured exudates which set into solid masses upon the surface of affected parts, common in stone fruits and citrus species.

(d) *Resinosis*, or the abnormal exudation of resin or pitch from coniferous species.

(e) *Latexosis*, or an abnormal outflow of latex, a milky fluid characteristic of certain species.

Rotting of Tissue

Succulent or woody stems and roots, fleshy leaves, flower buds or fruits may be affected with either dry or soft rot—the "gangrene" of plant tissue. The character of the rot may depend on the structures involved, the causal factors or complications and external conditions, but two types arc recognised; soft or watery rots and dry rots. Rots may be grouped according to the special structures or organs involved as:

(a) *Root rots*, illustrated by decay of such root crops as beets, carrots, parsnips, turnips, sweet potatoes, etc., the invasion of fleshy roots of such herbaceous crops as cotton or alfalfa and the destruction of woody roots of our forest, shade or fruit trees, and shrubs.

(b) *Leaf* or *stem rots*, illustrated by the decay of succulent or herbaceous leaves or stem as in the late blight of potatoes, the slimy soft rot of lettuce or the stem rot or wilt of various garden and field crops; by the decay of modified stems such as tubers, rhizomes, bulbs or corms; and by the decomposition of woody stems including live standing timber, dead timber and timber products such as posts, poles and lumber.

(c) *Bud rots* or the decay of fleshy buds of which the bud rot of carnations, the bud rot of the coconut and the black rot of the cabbage are notable examples.

(d) *Fruit rots*, which are soft or dry, varying with the nature of the fruit, the causal agent, and the environmental factors which are operative.

The great majority of rots are of parasitic origin and are induced by either bacteria or fungi. Rotting may, however, result from the operation of non-parasitic factors, as in the so-called blossom-end rot of the tomato, or the blossom-end rot of the watermelon. Rots of various kinds are the cause of heavy losses to growing crops or standing timber, and to vegetables or fruits while in storage or during transportation to market.

4

FIGHTING DISEASES

With the world's population continuing to rise, the major target of modern agriculture has to be a sustainable increase in yield that keeps pace with the increasing number of mouths to feed. This may be achieved by several approaches, not least by a significant reduction in pre- and postharvest losses due to disease. This chapter will describe some of the progress that has been made by plant biotechnology in combating plant diseases. It will briefly review the types of diseases, the organisms involved and the costs due to these diseases, and then move on to consider the mechanisms of disease resistance round in normal plants and how these are being enhanced using genetic manipulation.

Existing Non-GM Approaches

At present, many diseases are controlled by the large-scale use of expensive chemical treatments that kill either the pathogen or the vectors that carry them. At present $US700 000 000 is spent on fungicides in the USA annually. These chemicals may, by their very nature, be detrimental to human health and the environment. An example of this is the fumigant methyl bromide that has been used to control pathogens, insects and nematodes.In the USA, some 30000 tons of the chemical have been used each year to fumigate the soil prior to planting crops, to fumigate harvested crops during storage and prior to export and import (to kill any potential pathogens). Methyl bromide has been identified as a chemical that contributes to depletion of the ozone layer, and as such its use is now being phased out. This has left the agricultural industry with the task of developing methyl bromide-free farming practices. In this chapter, we deal with some examples of strategies being developed to reduce the use of these chemicals.

It is difficult to quantify the damage done by all the different plant diseases. It has been estimated that crop losses in the USA alone cost some $U533000000000 ($33 billion) per annum. In human terms, the loss of a crop by a subsistence farmer is as damaging, if not more so, than the loss of several thousand acres in the mid-west of the USA. These huge losses are partly due to the use of monocultures which encourage epidemics of pests and diseases. This is exemplified by the case of Southern corn leaf blight caused by the fungus *Bipolaris maydis*. Maize hybrids, the main form of the crop, are produced by breeding female, cytoplasmic male sterility lines (CMS) with male pollinators, under controlled conditions. By the late 1960s, maize varieties developed from the same CMS line (CMS-T) were being used in large areas of the USA. Unfortunately, this germ plasm was sensitive to the blight. Large areas of maize crop were decimated by what had previously been a minor form of leaf blight, but was now a very virulent form of the disease. Although this is an extreme case, it highlighted the problems of monocultures. Recent work has begun to explain why dispersed plant regimes are better than monocultures. One particular point, which is discussed later in this chapter in the section on inducible resistance, is that wild or mixed populations have a better chance of surviving attack from a virulent pathogen because they are likely to display polymorph isms in their disease resistance systems. Monocultures are less likely to do so. There are two other important points against monocultures: (1) that different plant types add to the diversity of environmental conditions, making them less beneficial for particular pathogens; and (2) that the increased distance between plant genotypes dilutes the inoculum of a given pathogenic race as it is dispersed between compatible host varieties.

Agricultural seed companies have, with some success, developed diseaseresistant lines. In the past this may have taken many years of intensive breeding programs to produce the plants, which may not necessarily have all the desired properties or the required level of resistance.

In the past, breeding strategies relied on basic observation of the phenotype in mature plants, but now they are being enhanced by the use of markers and molecular biology techniques. In the major crop plants, restriction fragment length polymorphisms (RFLP) and quantitative trait loci (QTL) maps, augmented with the results of the genome sequencing projects and PCR (polymerase chain reaction) - based molecular markers such as RAPD (random amplified polymorphic

DNA), AFLP (amplification fragment length polymorphism), microsatellites and STS (sequence tagged site) provide a battery of very powerful tools that allow genesltraits to be followed during the breeding process. Recently marker-based breeding has been used to 'pyramid' multiple disease resistance genes into plants. These combined approaches offer rapid and cleaner routes to broad-spectrum, durable disease resistance, which could lead to a significant improvement in yield with a concomitant reduction in chemical usage. Marker-assisted breeding strategies have a lot of merit. However, they depend upon the existence of resistance genes in the germplasm and the molecular information being available.

Transformation approaches have significant value in this area. To understand the approaches being used and the opportunities available we need to consider the range of plant diseases, the organisms involved and the natural mechanisms of disease resistance that can be optimised by gene manipulation.

Plant Pathogen Interactions

Interactions between plants and microorganisms can be viewed simply as being of four different types:

1. The microorganisms form symbiotic relationships with the host, these include bacteria like *Rhizobium* spp. and the rhizo-fungi.
2. The microorganism may cause disease in the host plant.
3. The host plant may be resistant to the pathogen, and no infection develops.
4. The host plant shows some tolerance to infection, the pathogen is able to grow and replicate but symptoms of infection are minimal.

Within this broad sweep of interactions is a whole range of others that are now beginning to be understood at the molecular level. The pathogens themselves are of two basic types: (1) necrotrophs, i.e. organisms that kill the host and feed on the contents, and which are often associated with toxin production; and (2) biotrophs, i.e. those that require a living host to complete their life cycle. Bacteria and fungi may be of either type, whilst viruses are obligate biotrophs. In the following sections a brief summary of the main types of organisms and their effects will be given.

Prokaryotes

Amongst the prokaryotes, the most economically important pathogenic organisms are members of the phytoplasma group (previously known as mycoplasma-like organisms) and o.rganisms from a small number of bacterial genera. The phytoplasma belong to a group of prokaryotes that lack cell walls and survive as either saprophytes (organisms that live on dead organic matter which they help to break down) or as intracellular pathogens. Diseases include aster yellows, in which the phytoplasma is spread by the aster leafhopper. The pathogen infects over 300 host plant species, including flax, canola, wheat, barley and potato. Killing the leafhopper vector with chemicals controls the disease.

The main groups of bacterial plant pathogens include members of the genus *Agrobacterium*, which cause crown galls and hairy roots by transferring T-DNA into plant cells. Some pathogens can enter the host through natural openings such as stomata, but, like many other pathogens, *Agrobacterium* enters the plant through wound sites. Other bacterial pathogens are found amongst the *Corynebacterium*, *Erwinia*, *Pseudomonas*, *Xanthomonas* and the *Streptomyces* genera. Many of the symptoms caused by the different infestations of bacteria are similar: spots, galls, cankers (localised death of an organ), blights (rapid discoloration, wilting, death), soft rots (caused by secreted enzymes that break down plant cell walls and reduce storage time) and wilts (caused by bacteria blocking the vascular tissues). These infections can be localised and have a minimum commercial effect, disfiguring the fruit or leaves, or they can have major effects on yield or commercial value. For instance, the bacterial disease fire blight, caused by *Erwinia* spp. can kill whole fruit trees in a single season.

Fungi and Water Moulds

The vast majority of cellular pathogens of plants are found amongst the fungi. Whilst not directly affecting human health, the losses of food caused by fungal diseases have had profound effects on humanity. There are over 100000 known species of fungi. Most are examples of saprophytes, but the fungi also include symbionts, such as the mycorrhiza, and perhaps more than 8000 species that cause plant drseases. Modes of entry include wound sites and natural openings such as stomata, and the enzymes that break down dead organic matter can be used by pathogens to gain entry into living plants by degrading cell wall macromolecules. There are too many fungal pathogens to go into any detail of the diseases they cause or their modes of action. *Phytophthora infestans* (which is a classic example of an Oomycete-group pathogen), the organism responsible for the potato blight that caused the great Irish famine of 1846. Until recently, Oomycetes have been classified as fungi; however, biochemical and molecular studies indicate that these 'water moulds' should now be classified separately. The cell walls of these organisms are composed of cellulosic compounds and glycan not chitin, and the nuclei within the hypha (filaments) are diploid, not haploid, as in fungi.

Viruses

It is important to note that viruses constitute a major group of pathogens, and that many of the natural host defences discussed below are induced by the cellular and viral pathogens.

Some effort has been made in recent years to classify plant viruses (families, genus, species, etc.). Full details of 950 plant viruses, their hosts and the diseases they cause can be found at Plant Viruses Online. The vast majority of plant viruses are single-stranded, positive-sense RNA viruses, but other types such as circular double-stranded DNA viruses exist. Interestingly, one of the tungro virus disease, often involves two viruses: rice tungro bacilliform *badnavirus* (a DNA virus) and rice tungro spherical *waikavirus* (an RNA virus). This *badnavirus* causes the tungro symptoms (yellow-orange leaves and stunting), but it depends on the *waikavirus* for transmission by leafhoppers (the *waikavirus* also increases the severity of the tungro symptoms), whereas the *waikavirus* alone causes mild stunting.

Natural Disease Resistance Pathways—Overlap between Pests and Diseases

It would be wrong to give the impression that plants have no resistance against pathogen attack, for it is clear they do. They do not have an immune system that produces specific cells to attack invading microbes, rather they have adopted more general defence systems. There is also a lot of overlap between the plant's response to pathogens and plant pests. The general cellular damage caused by both can act as a signal to trigger general defence systems. There are advantages to the plant having such general systems because pests often act as vectors for pathogens. As we shall see below, plants also have systems that respond to specific signals. However, much of the recent molecular work has been done with model plants such as tobacco and *Arabidopsis* and it is possible that the observations are not applicable to all plants. In this section, four different levels of defence will be considered:

1. anatomical defences;
2. pre-existing protein and chemical protection;
3. inducible systems;
4. systemic responses.

Anatomical Defences

As indicated earlier, many pathogens have to invade plants through wounds. This is because plants have developed morphological and structural systems that preclude pathogen access to living cells (the first line of defence). These can be thick layers of cuticle, bark, waxes, etc. Once this defence is breached then cascades of defence systems come into play.

Pre-existing Protein and Chemical Protection

The second line of defence is made up of a range of antimicrobial proteins produced by the plant during growth and development, these include defensins and defensin-like proteins. The defensin proteins are similar to those found In insects and mammals where they play an important role in defence against infectious agents. Their structure has a conserved three-dimensional folding pattern, which suggests they represent a superfamily of peptides that pre-dates the divergence of plants and animals. Some defensins cause increased branching in fungi, while others simply slow growth. They are frequently associated with seeds at the time of germination, when they may be released into the environment and create a microenvironment around the seed that suppresses fungal growth. Many of the large number of small chemicals made as secondary products may also have antimicrobial properties. These proteins and chemicals may simply deter pest or pathogen growth, or they may actually be toxic to them.

Inducible Systems

The third level of defence is a switch to counterattack that relies on *de novo* protein synthesis. It would be costly for the plant to have its defence system permanently switched on, so there are mechanisms in place to detect the infection and then turn on the defence system. These mechanisms will be discussed below.

When a pathogen arrives at its host and gains entry to living cells it may induce resistance to infection. This response can be divided into three parts. First, there may be a local response that involves interactions with molecules released by the pathogen (elicitors). The second, recognition-dependent disease resistance, is based upon an interaction between specific proteins produced by the pathogen (the avirulence gene product) and a protein produced by the plant (the resistance gene products). Both these interactions may lead to a cascade of reactions that invoke the hypersensitive response (HR). The third part of the response is the induction of a systemic resistance (and even the passage of signals to other plants).

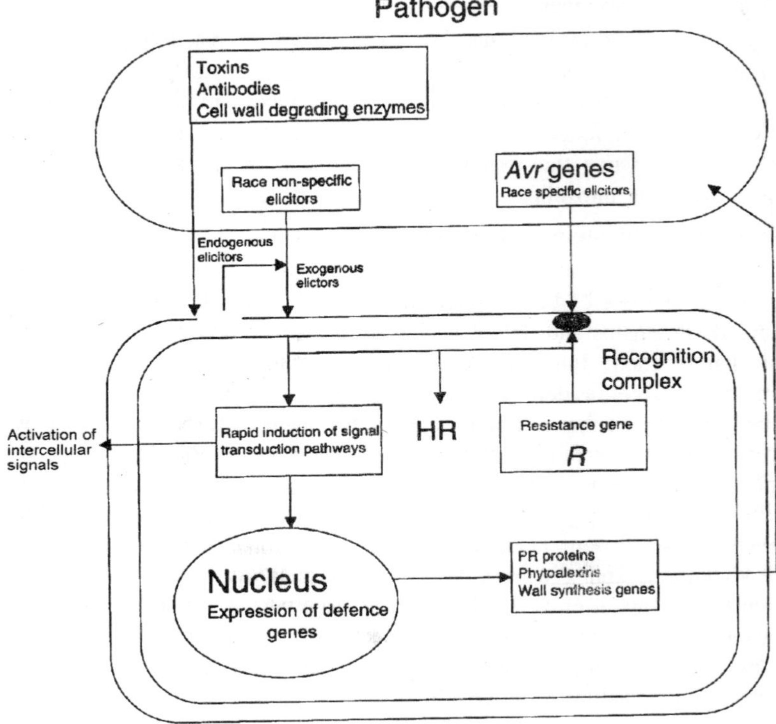

Fig. 4.1. Cellular response to pathogen attack.

Elicitor response

The initial local reaction has several parts. In the first instance there is normally damage to the plant cell wall, causing the release of wall fragments such as pectic-oligomers. These act as signals—endogenous elicitors—which bind to specific receptors, setting off a cascade of reactions that lead to the induction of specific defence genes. These defence genes code for enzymes that synthesise structural components for cell wall thickening (to repair the damage), enzymes of secondary metabolism, lectins (multimeric sugar-binding proteins that agglutinate cells) and many so-called 'pathogenesis-related (PR) proteins'. PR proteins include chitinases and β-1,3-glucanases, protease inhibitors, non-specific lipid transfer proteins, ribosomal inhibitor proteins and various antimicrobial proteins. These antimicrobial proteins include defensins such as SN1, which is active against bacterial and fungal pathogens in potato. One important point of relevance to the use of the genes in transgenic experiments, is that a number of the PR gene products have been identified as latex allergens. This is a family of proteins found in many plant species and responsible for serious medical conditions in many people who show allergenic response to latex products and a number of fruit species.

The effect of synthesising the defence proteins depends, in part, on the pathogen. For instance, if fungal hyphae breach the cells defences then the chitanases and glucanases may cause some degradation of the pathogen's cell wall. This will lead to the production of chitin and β-1,3-glucan oligomers. These compounds may act as signal molecules (exogenous elicitors) which bind to membrane receptors and re-enforce the induction of the defence systems. The cells may also produce phytoalexins (phenolic compounds or terpenes) that kill any pathogens as well as the cells in the vicinity of the infection, therefore limiting the spread.

Recognition-dependent disease resistance

Using a genetic analysis of the interactions between flax and flax rust, Harold Flor developed the gene-for-gene hypothesis more than 50 years ago. The idea stems from the observation that disease resistance requires two complementary genes. The pathogen carries the avirulence

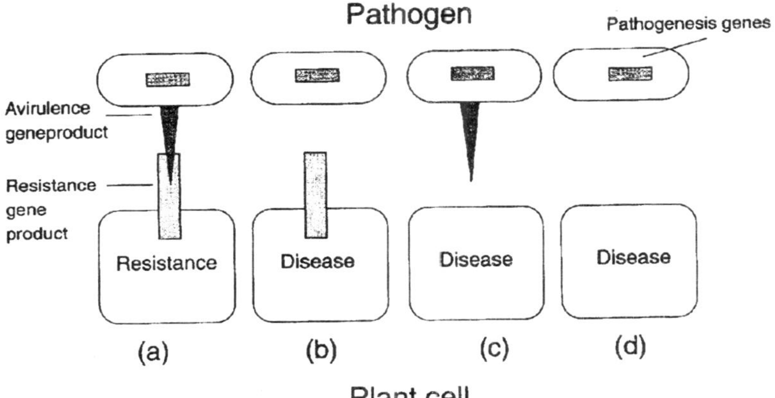

Fig. 4.2. HR interactions In (A) the plant cell contains the resistance gene (R) and the pathogen produces the avirulence gene product (Avr); the plant shows resistant of infection. In (B) the pathogen is not producing the Avr protein so the interaction results in disease. In (C) host cell is not producing the R protein so disease results. In (D) neither the R protein nor the Avir protein is being produced so disease results.

gene (*Avr*) which codes for a protein that is recognised by a specific receptor protein in the plant cell—encoded for by the resistance gene (*R*). This interaction induces the hypersensitive response (HR), which is manifested as a local necrosis that develops through a mitochondrial-associated NADPH-dependent oxidative burst and/or the release of phenolics and nitric oxide. The activation of signalling pathways also leads to the induction of many of the pathogenesis-related proteins. The most important feature of this recognition system is that if either of the proteins is absent then the pathogen will cause disease in the plant. In panel (a) both proteins are present so the defence systems are activated and the plant is resistant to the pathogen. In panel (b) the avirulence gene is absent, a local HR reaction is not induced and the plant develops the disease. In panel (c) the plant is missing the R gene, and in (d) both proteins are missing. Again the plant develops the disease in both cases.

The gene-gene system is one of the most important ways plants have of switching on resistance systems, and it has been shown to function for interactions between plants and aphids, nematodes, fungi, viruses and bacteria. Although little is known about the Avr proteins, it is postulated they might function as virulence factors, subverting cellular functions through interactions with plant-encoded pathogenicity targets. Many examples of the R proteins have now been identified: the annotation of the *Arabidopsis* genome sequence has indicated there are about 100 R loci distributed through the genome. They fit into

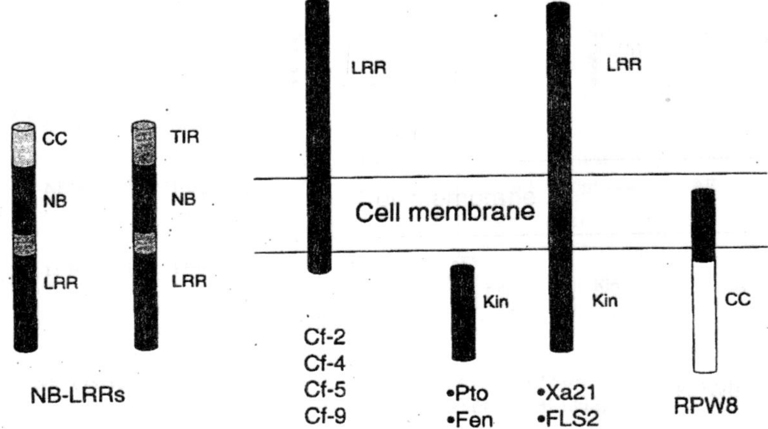

Fig. 4.3. A representation of the location and structure of the five main classes of plant disease proteins.

five basic structural groups. In the biggest group, the proteins are characterised by having a nucleotide-binding site and a leucine-rich repeat region (these are termed 'NB-LRR proteins'). The plant R proteins show extensive similarity with mammalian and insect receptor proteins required for the onset of the innate immune response and that are involved in the sensing of pathogen-derived factors called 'PAMPs' (for pathogen-associated molecular patterns).

The other four groups are structurally diverse, but examples have been found in which the R proteins also contain an LRR region or, as will be discussed below, the R protein interacts with another protein that contains the LRR domain. It is intriguing to consider that such a small number of R proteins can give the diversity required to deal with the large number of plant pathogens, and how they evolve to deal with evolutionary changes in the Avr protein. One factor seeming to favour plant defence mechanisms is that Avr proteins are key factors in the biology of the pathogen, and as such that they do not evolve at high rates. There is also evidence indicating that to meet with the changes that do occur, some variability in the plant R proteins comes from the fact that the LRR domain does mutate, within a highly variable region, and that mutations encoding new amino acids are selected for. The R-gene loci are very polymorphic in wild populations, so with each R-gene allele being present at a low frequency there is limited selection for virulence in the pathogen population. In conventional agriculture, with the use of mono cultures, the balance between polymorphisms within the resistance genes is disrupted therefore making the crops potentially more susceptible to virulent infections. The mechanisms by which interactions between Rand Avr bring about the HR response and induce the expression of the defence proteins are now being uncovered.

Systemic Responses

The induction of local defence pathways may lead to the induction of intercellular signals that produce a systemic response, termed 'systemic acquired resistance' (SAR). Both avirulent and virulent pathogens may result in the induction of SAR, although it is usually a slower process in the case of virulent pathogens. SAR has two phases: the initiation phase and the maintenance phase. In the initiation phase, cells at the foci of the infection release signal molecules, typically salicylic acid (SA), into the phloem. These are transported to target cells in other parts of the plant where SAR genes (such as certain PR proteins, etc.) are expressed, thus giving the plant some level of

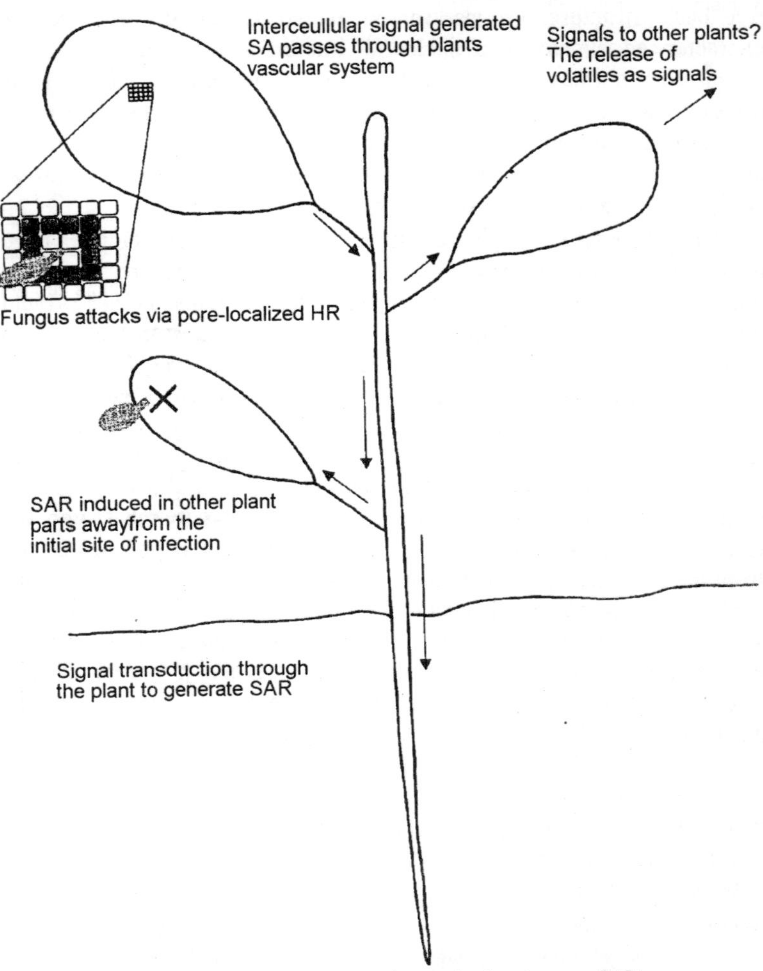

Fig. 4.4. Induction of systemic acquired resistance (SAR).

resistance against infection. In the longer term, a maintenance phase is reached that may last for weeks or even the full life of the plant, in which there is a quasi steady state resistance against virulent pathogens.

Other states of disease resis tance, such as induced systemic resistance (ISR), which are independent of salicylic acid, are also induced by avirulent pathogens through pathways that seem to include ethylene or jasmonic acid as messengers. These may result in the induction of different classes of defence proteins. Evidence now suggests that signals carried by volatile chemicals, such as methyl-jasmonate,

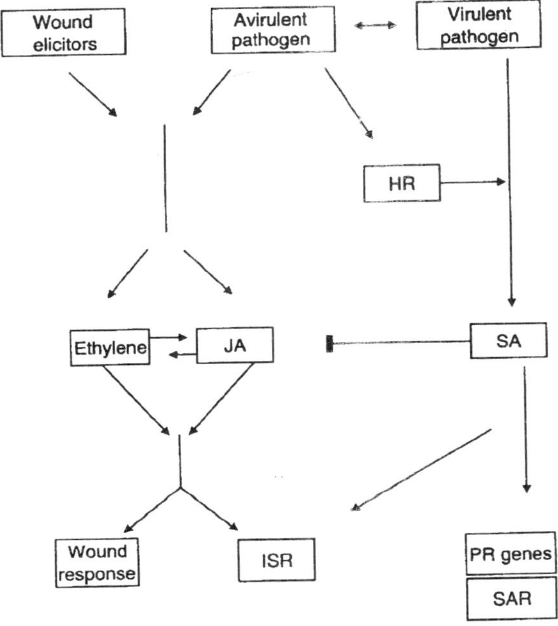

Fig. 4.5. Network of disease response pathways.

can travel between plants to warn of an attack. Many of the proteins involved in these pathways have been identified and are providing further insights into disease resistance.

Biotechnological Approaches to Disease Resistance

A great deal of progress has been made in converting the fundamental information obtained on resistance into strategies for enhancing plant defence systems. As one would imagine, these range from strategies in which individual pathogens are targeted to approaches designed to enhance general resistance systems to inhibit a range of pathogens. Many have been tested on model systems, some have been introduced into crop plants, but none have so far been commercialised. Several of the systems used to genetically engineer plant resistance to bacterial and fungal pathogens will be considered in more detail in the following section of this chapter.

Protection against Fungal Pathogens

Initial studies for the enhancement of resistance have been based upon the plant's own defence systems, in particular the transformation of plants with genes encoding for PR proteins. To combat fungal pathogens, many plants have been transformed with genes that code for chitinase and glucanase enzymes. These degrade polymers in the cell walls of many, but not all, fungi, without affecting the host plant or any animals. The genes for these enzymes have been isolated from a number of sources, including plants (rice, barley), bacteria (*Serratia marcescens*) and even fungi (*Trichoderma harzianum*). The choice of enzyme used has partly depended on its availability and on its effectiveness against the pathogen being addressed, since successful pathogens are already likely to have resistance against plant systems. The 355 promoter has often been used to express these genes. Constitutive expression of the defence protein is designed to provide a barrier against the initial attack and not allow the pathogen to establish, as may be the case during the induction period of the host defence system. Wound inducible promoters such as the potato prp1-1 promoter have also been used, as they mediate rapid and localised expression in response to pathogen attack. This type of promoter restricts gene expression to where the protein is required and may reduce any potential hazards associated with the proteins. As gene expression is only turned on during the infection therefore limiting gene expression, it means there is no serious drain on the plant's biosynthetic capability nor a serious amount of yield reduction (yield drag).

The efficiency of these systems has been tested in many plants, and in many cases some level of resistance/tolerance to specific fungal pathogens has been demonstrated. One of the earliest demonstrations of tolerance was exhibited by field-grown transgenic tobacco lines, containing a chitinase gene from the bacterium *Serratia marcescens*. Both intracellular and extracellular forms of the enzyme (glycosylated or unglycosylated forms) gave the plants increased tolerance to the pathogen *Rhizoctonia solani*. In another example, transgenic cucumber plants showed a range of responses to the grey mould (*Botrytis cinerea*) when transformed with a rice chitinase sequence (RCC2) driven by a 355 promoter. Interestingly, high-level resistance was observed in a number of lines where spread of the disease was blocked completely. However, the block in fungal attack occurred at different stages in the infection process. Some lines prevented hyphal penetration, whilst others allowed penetration but restricted the spread.

Perhaps the most ambitious work with chitinase genes has been carried out with fruit trees. Apple tree plants expressing either an exochitinase and/or an endochitinase isolated from the fungus *Trichoderma harzianum* have been produced using *Agrobacterium*-

mediated transformation. A positive correlation was found to exist between resistance to scab produced by the fungus *Venturia inaequalis* and the level of expression of the proteins. If both genes were introduced into plants the proteins acted synergistically. One drawback was that expression of the endochitinase reduced plant growth. One important question that remains to be answered is how long the resistance will be maintained. A further disadvantage of the chitinase approach is that major Oomycete pathogens such as *Phytophthora* spp. do not have chitin in their cell walls. To overcome these difficulties, combinations of PR proteins have been used and other antimicrobial proteins have been tested. This is a sensible strategy anyway, as the use of more than one gene will help to prevent the development of resistance and provide longer term protection.

Glucanases, which are also one of the groups of PR proteins, have been used as a tool to enhance resistance to fungal infection. There are different forms of the enzyme with different substrate specificities, but one of the most widely used is a class-II β-1,3-glucanase from barley. When expressed in transgenic tobacco plants under control of the 35S promoter, the gene provided increased protection against the soil-borne fungal pathogen *Rhizoctonia solani*. In plants also expressing chitinase transgenes, there is a synergistic effect and resistance is further enhanced. A third type of protein, ribosome inactivating proteins (RIP), has also been used as a defence against fungal infections. RIPs are enzymes that remove an adenine residue from a specific site in the large rRNA of eukaryote and prokaryote ribosomes, thereby inhibiting protein synthesis. The specificities of these proteins vary, but the critical point is that certain examples do not inhibit plant ribosomes. Type-1 barley RIP, expressed constitutively, has been used to provide resistance to fungal attack and has been tested in tandem with the chitinase gene. Again, a synergistic effect was found against *Rhizoctonia solani* in transgenic tobacco plants.

Transgenic crops for food safety

Recently, transgenic approaches have been used to combat the effects of fungal infection rather-than the infection itself. *Fusarium verticillioides* causes fusarium ear mould in maize and produces fumonisin mycotoxins, which are acutely toxic to certain livestock and are also carcinogenic. It has not yet been possible to incorporate ear-mould resistance into the maize crop, but it has been possible to incorporate either a fungal esterase gene or a fungal amine oxidase gene into maize and detoxify the maize to some extent.

Antimicrobial Proteins

Many other genes that code for proteins with both antifungal and antibacterial activity have been introduced into plants with varying success. The proteins used have included lysozyme (this enzyme degrades chitin as well as peptidoglycan) which delayed fungal infection for a short time. Potato plants expressing lysozyme have been studied with respect to their resistance to *Erwinia carotovora*. Various thionin proteins have been introduced into plants with mixed success, barley a-thionin expressed in transgenic tobacco was shown to increase resistance to *Pseudomonas syringae*, but other attempts did not enhance resistance.

An exciting area of development is in the area of defensins. These proteins are found in all living cells. They have various properties, but are small antimicrobial peptides (26-50 amino acid residues) that are ancient mediators of innate defences. Defensins display lytic activity by binding within microbial plasma membranes. This sort of attack may prove difficult for the pathogen to develop resistance to. As the integrity of the membrane is critical, it is unlikely that changes in the structure of the membrane could occur to prevent defensin infiltration. An alfalfa defensin gene, *alfAFP,* has been introduced into potatoes and its performance monitored in the field. The defensin was processed and secreted into extracellular spaces of leaves and roots in transgenic plants. In the greenhouse and in several field trials, the transgenic plants were shown to have a significant resistance toward the fungus *Verticillium dahlia*. The level of resistance was comparable to that found with a naturally resistant potato. However, they were also expected to be resistant to *Alternaria solani*, but this proved not to be the case.

An artificial defensin gene has been made that incorporates the *cecropin* and *melittin* genes, obtained from a giant silk moth and bee, respectively. This gene has also been introduced into potatoes, and transgenic plants challenged in tissue culture conditions with the bacterium *Erwinia carotovora*. Tubers were then tested for *E. carotovora* soft rot—during the course of the experiment the transgenic tubers remained firm, whereas control plants rotted. This type of study also shows the power of transgenic approaches where genes with the required characters can be moved across kingdoms.

Some of the most impressive studies in this area have been carried out to combat the bacterial disease fireblight. Many of the most important commercial apple cultivars and rootstocks grown in Europe are sensitive to fireblight. This disease is caused by *Erwinia amylovora* which, once established, can kill a tree within a growing season. It is also a very serious disease of pears. A number of independent groups are addressing this problem by producing transgenic trees containing the genes for antimicrobial proteins. One strategy used has been the 'lytic' approach, to produce transgenic fruit trees (via *Agrobacterium-mediated* transformation) containing the gene for T4 lysozyme and genes for the insect antimicrobial proteins attacin E and cecropin. A second approach being used involves the inclusion of a bovine lactoferrin gene to compete with microbial siderophores and a depolymerase gene whose product is capable of degrading specific exopolysaccharides.

These experiments are at various stages, but field trials have been carried out with apple plants. It is claimed that plants expressing attacin E under the control of a constitutive promoter show increased field resistance to fire blight. Experiments were carried out in 1998 in which 2- and 3-year-old plants were inoculated with *E. amylovora* and the percentage of the current season's shoot-length blighted (SLB) tissue determined. Many of the transgenic lines showed less blight than the control plants. One line in particular had only 5% SLB compared with 56% in non-transgenic controls.

The examples given above highlight the use of antimicrobial proteins to interact directly with pathogens and stop infection. One can ask, though, will they ever be commercialised? It is likely some will, but so far these approaches have been disappointing, as even when using non-specific enzymes the spectrum of resistance achieved has been limited. This may be a problem, as one view is that a broad-spectrum of resistance in the plant lines is more likely to be acceptable to the farmer than if only single pathogens are being dealt with. The alternative view, as exemplified by the approach to fire blight, is that some diseases are so important that strategies that deal with these alone are acceptable. Other strategies have been tried in order to introduce broad-spectrum resistance; for instance, strategies to switch on resistance pathways have been developed.

Induction of HR and SAR in Transgenic Plants

From the information given above, it is clear that the most obvious approach for switching on a general resistance pathway would be to use the gene-togene interaction system. One possible implication of the Guard hypothesis is that the R protein has a function in the cell that relates to the general resistance mechanism. This would suggest that an increase in the production of the R protein will bring about an increase in the expression of the resistance systems. This hypothesis has been tested in a series of experiments in which the *pto* gene, under the control of the 355 promoter, has been introduced into tomato plants.

Close examination of three transgenic lines indicated that over expression of the Pto protein activated a defence response in the absence of the Pto-AvrPto interaction. Cell death due to an HR response was highly localised to palisade mesophyll cells. The mesophyll cells also showed accumulation of autofluorescent compounds, callose deposition and lignification. Leaves also exhibited salicylic acid accumulation and increased expression of pathogenesis-related genes. The important outcome of this work is that this constitutive expression of the resistance systems is not species-specific. Plants that normally would be sensitive to *Pseudomonas syringae pv tomato*, without *avrPto,* were shown to be resistant. They were also tested with *Xanthomonas campestris pv vesicatoria*, a leaf pathogen that causes bacterial spot disease and *Cladosporium fulvum*, which causes leaf-mould disease. The transgenic plants were also more resistant to these pathogens.

Another approach that has been adopted to switch on the HR-related defence system has been devised, in which a *Phytophthora cryptogea* gene that codes for a highly active elicitor termed 'crytogein' has been fused to a pathogen-inducible promoter from tobacco and introduced into transgenic tobacco. Under non-induced conditions (i.e. in the absence of the pathogen) no elicitor is made. On infection by the virulent fungus *Phytophthora parastica var nicotianae* the crytogein gene was expressed, and the elicitor-induced defence gene expression was detected around the infection site. Localised necrosis, similar to what would have been produced by non-virulent pathogens, occurred around the infection site. The plants also exhibited resistance to other fungi unrelated to the *Phytophthora* genus. Broad-spectrum resistance, at least to fungi, has been produced without the constitutive expression of the transgene.

5

PHYSICAL FACTORS

Plants, like human beings or animals, grow best within certain ranges of the various factors that make up their environment. Such factors include temperature, soil moisture, light, air composition and contaminants, solid composition and pH. Although these factors affect all plants growing nature, their importance is considerably greater for cultivated plants—which are often grown by man in areas barely meeting the requirements for normal growth of the particular plants. Moreover, cultivated plants are frequently grown or kept in completely artificial environmental (greenhouses, warehouses, etc.) or a subjected a to a number of cultural practices (fertilisation, irrigation, spraying with pesticides, etc.) which may affect their growth considerably. The common characteristic of non-infectious diseases of plants is that they are caused by lack or excess of any of the conditions supporting life or by an agent that interferes with the supply thereof.

Non-infectious diseases occur in the absence of pathogens, and cannot, therefore, be transmitted from diseased to healthy plants. Non-infectious diseases may affect plants in any or all of their life stages, such as seed, seedling, mature plant, or fruit, and they may cause damage in the field, storage, or market. The symptoms caused by noninfectious diseases vary in kind and severity with the particular environmental factor involved and with the degree of deviation of this factor from its range which supports normal plant growth. Symptoms may range from slight to severe, and affected plants may even die. The diagnosis of non-infectious diseases is sometimes made easy by the presence on the plant of characteristic symptoms known to be caused by the lack or excess of a particular factor. At other times diagnosis can be arrived at by carefully examining and analyzing the weather conditions prevailing before and during the appearance of the disease, recent changes in the atmospheric and soil contaminants at or near the area where the plants are growing, and the cultural practices, or possible accidents in the course of these practices, preceding the appearance of the disease.

Often, however, the symptoms of several non-infectious diseases are too indistinctive and closely resemble those caused by several viruses and the many root pathogens. The diagnosis of such non-infectious diseases then becomes a great deal more complicated and depends on proof of absence from the plant of any of the pathogens that could cause the disease, and on reproducing the disease on healthy plants after subjecting them to conditions similar to those thought of as the cause of the disease. To distinguish further among environmental factors causing similar symptoms, the investigator must cure the diseased plants, if possible, by growing them under conditions in which the degree or the amount of the suspected environmental factor involved has been adjusted to normal. Noninfectious plant diseases can be controlled by avoiding the extremes of the environmental conditions responsible for such diseases, or by supplying the plants with protection or substances that would bring these conditions to levels favourable for plant growth.

Temperature

Plants normally grow at a temperature range from 1 to 40^{0}C, most kinds of plants growing best between 15^0 and 30^{0}C. Perennial plants and dormant organs, such as seeds and corms, of annual plants may survive temperatures considerably below or above the normal temperature range of 1^0 - 40^{0}C. Vegetative growth, however, especially the young, growing tissues of most plants and the entire growth of many annual plants, is usually very sensitive to temperature near or beyond the extremes of this range. The minimum and maximum temperatures at which plants can still produce normal growth vary greatly with the plant species and with the stage of growth the plant is in during the low or high temperatures. Thus, plants such as tomato, citrus, and other tropical plants grow best at high temperatures and are injured severaly when the temperature drops to near, or below, the freezing point.

On the othe hand, plants such as cabbage, winter wheat, alfalfa, and most perennials of the temperate zone can withstand temperatures considerably below freezing without any apparent ill effects to the plant. Even the latter plants, however, will be injured and finally killed if the temperature drops too low. A plant may also differ in its ability to withstand extremes in temperature at different stages of its growth. Thus, older, hardened plants are more resistant to low temperatures than are young seedlings. Also, different tissues or organs on the same plant may vary greatly in their sensitivity to the same low temperature. Buds are more sensitive than twigs, flowers and newly formed fruit are more sensitive than leaves, and so on. Plants are generally injured faster and to a greater extent when temperatures become higher than the maximum for plant growth than when they are lower than the minimum. However, too high a temperature rarely occurs in nature and, therefore, few disorders of significance can be attributed to too high temperatures.

Even in the best documented cases, high temperature seems to cause its effects on the plant in conjunction with the effects of other environmental factors particularly excessive light, drought, lack of oxygen, or high winds accompanied by low relative humidity. High temperatures are usually responsible for sunscald injuries appearing on the sun-exposed sides of fleshy fruits and vegetables, such as apples, tomatoes, onion bulbs, and potato tubers. On hot, sunny days the temperature of the fruit tissues beneath the surface facing the sun may be much higher than that of those on the shaded side and of the surrounding air. This results in discolouration, water-soaked appearance, blistering, and a desiccation of the tissues beneath the skin which leads to sunken areas on the fruit surface.

Succulent leaves of plants may also develop sunscald symptoms, especially when hot sunny days follows periods of cloudy, rainy weather. Irregular areas on the leaves become pale green at first but soon collapse and form brown, dry spots. Too high a soil temperature at the soil line sometimes kills young seedlings or causes cankers at the crown on the stems of older plants. High temperatures also seem to be involved in the water core disorder of apples and, in combination with reduced oxygen, in the blackheart of potatoes. Far greater damage to crops is caused by low than by high temperatures.

Low temperatures, even if above freezing, may damage warm-weather plants such as corn and beans. They may also cause excessive sweetening and, upon frying, undesirable caramelisation of potatoes due to hydrolysis of starch to sugars at the low temperatures. Temperatures below freezing cause a variety of injuries to plants. Such injuries include the damage caused by late frosts to young meristematic tips or entire herbaceous plants, the frost-killing of buds of peach, cherry, and other trees, and the killing of flowers, young fruit, and, sometimes, succulent twigs of most trees. Low winter temperatures may kill young roots of trees, such as apple, and may also cause bark-splitting and canker development on trunks and large branches, especially on the sun-exposed side, of several kinds of fruit trees. Fleshy tissues, such as potato tubers, may be injured at sub-freezing temperatures. The injury varies depending on the degree of the temperature drop or the duration of the low temperature.

Injury only on the main vascular tissues appears as a ringlike necrosis, injury of the finer vascular elements which are interspersed in the tuber gives the appearance of netlike necrosis, and with more general injury large chunks of the tuber are damaged creating the so-called "blotch-type" necrosis. The mechanisms by which high and low temperature injure plants are quite different. High temperatures apparently act by inactivating certain enzyme systems and accelerating others, thus leading to abnormal biochemical reactions and death of the cell.

High temperature may also cause coagulation and denaturation of proteins, disruption of cytoplasmic membranes, suffocation, and possibly release of toxic products into the cell. The end effects of high temperature on plant cells and tissues depend on the maximum temperature and its duration. They may vary from a minor temporary disturbance in the physiology of the cells to death and desiccation. Low temperatures, on the other hand, injure plants primarily by inducing ice formation between and/or within the cells. The rather pure water of the intercellular spaces freezes first and at about 0^{0}C, while the water of the intercellular spaces freezes first and at about 0^{0}C, while the water within the cell contains solutes which, depending on their nature and concentration, may depress the freezing point of water for several degrees.

Further more, when the intercellular water becomes ice, the vapor pressure between cells decreases and more vapor (water) moves out of the cells and into the intercellular spaces, where it also becomes ice. The reduced water content of the cells depresses further the freezing point of the intracellular water and this could continue, up to a point, without damaging the cell. Below that point, however, ice crystals may form within the cell disrupting the plasma membrane and other organelles and systems and causing injury and death to the cell. The freezing point of water in cells varies with the tissues of the plant and with the species of plants; in some tissues of the winter-hardy species of the north, ice probably never forms within the cells regardless of how low the temperature become. Even when ice forms only in the intercellular spaces, cells and tissues may be damaged either by the inward pressure exerted by the ice crystals, or by loss of water from their protoplasm to the intercellular spaces. This loss causes plasmolysis and dehydration of the protoplasm, which may cause salting out or coagulation.

The rapidity of the temperature drop in a tissue is also important, since this affects the amount of water remaining in cell and, therefore, the freezing point of the cell contents. Thus, a rapid drop in temperature may result in intracellular ice formation where a slow drop to the same low temperature would not. The rate of thawing may have similarly variable effects, since rapid thawing may flood the area between cell wall and protoplast and may cause tearing and disruption of the protoplast if the latter is incapable of absorbing the water as far as it becomes available from the melting of ice in the intercellular spaces.

Moisture

Moisture disturbances in the soil are probably responsible for more plants growing poorly and being unproductive annually, over large areas, than any other single environmental factor. Small or large territories may suffer from drought over periods of time. The subnormal amounts of water available to plants in these areas may result in reduced growth, diseased appearance, or even death of the plants. Lack of moisture may also be localized in certain types of soil, slopes or thin soil layers underlaid by rock or sand and may result in patches of diseased-looking plants, while immediate surrounding areas appear to contain sufficient amounts of moisture and the plants in them grow normally. Plants suffering from lack of sufficient soil moisture usually remain stunted, are pale green to light yellow, have few, small and drooping leaves, flower and fruit sparingly and, if the drought continues, may wilt and die.

Although annual plants are considerably more susceptible to short periods of insufficient moisture, even perennial plants and trees may be damaged by prolonged periods of drought and may produce less growth, small scorched leaves and short twigs, die-lack, defoliation, and finally wilting and death. Plants weakened by drought are also rendered more susceptible than before to certain pathogens. Lack of moisture in the atmosphere, i.e., low relative humidity, is usually temporary and seldom causes damage. When combined with high wind velocity and high temperature, however, it may lead to excessive loss of water from the foliage and may result in leaf scorching or burning, and temporary or permanent wilting plants. Excessive soil moisture occurs much less often than drought where plants are grown; but poor drainage of flooding of planted fields may result in more serious and quicker damage, or death, to plants than that from lack of moisture.

Flooding during the growth season may cause permanent wilting and death of succulent annuals within 2-3 days. Trees, too, are killed by water logging, but the damage usually appears more slowly and after their roots have been continually flooded for several weeks. As a result of excessive soil moisture the fibrous roots of plants decay. Their underground storage organs collapse and are invaded by soft rot microorganism. The death of flooded roots is probably due both to the reduce supply of oxygen to the roots and to changes in the soil microflora brought about the excess water in and exclusion of atmospheric oxygen from the soil. Oxygen deprivation causes stress, asphyxiation, and collapse of many root cells.

Wet, anaerobic conditions favour the growth of anaerobic microorganisms which, during their life processes, form substances, such as nitrites, that are toxic to plants. Besides, the root cells damaged directly by the lack of oxygen lose their selective permeability and may allow toxic metals, etc., to be taken up by the plant. Also, once part of roots are killed, more damage is done by facultative parasites which may be greatly favoured by the new environment. Thus, the wilting of the plants which soon follows flooding, is probably the result of lack of water in the above ground parts of plants caused by the death of the roots, although it appears that translocated toxic substances may also be involved.

Inadequate Oxygen

Low oxygen conditions in nature are generally associated with high soil moisture and/or high temperatures. Lack of oxygen may cause desiccation of roots of different kinds of plants in waterlogged soils, as was mentioned under moisture effects. Root collapse of alfalfa has been shown to result from a combination of high soil moisture and high soil temperature, and to be influenced by the stage of growth of the plant after clipping. The first condition, apparently, reduces the amount of oxygen available to the roots while the other two increase the amount of oxygen required by the plants. The two effects together result in an extreme lack of oxygen in the roots and cause their collapse and death. Low oxygen levels may also occur in the centers of fleshy fruit or vegetables in the field, especially during period of rapid respiration at high temperatures, or in storage of these products in fairly bulky piles.

The best known such case is the development of the so-called blackheart of potato, in which fairly high temperatures stimulate respiration and abnormal enzymatic reactions in the potato tuber. The oxygen supply of the cells in the interior of the tuber is insufficient to sustain the increased respiration, and the cells die of suboxidation. Enzymatic reactions activated by the high temperature and suboxidation go on before, during, and after the death of the cells. These reactions abnormally oxidize normal plant constituents into dark melanin pigments. The pigments spread into the surrounding tuber tissues and finally make them appear black.

Light

Lack of sufficient light retards chlorophyll formation and promotes slender growth with long internodes and abnormal internal tissue. This condition is known as etiolation. Etiolated plants are found outdoors only when plants are spaced too close together or when they are growing under trees or other objects. Etiolation of various degrees, however, is rather common in greenhouses, seedbeds, and cold frames, where plants often receive inadequate light. Etiolated plants are usually thin and tall and are susceptible to lodging. Excess light is rather rare in nature and seldom injures plants.

The quality of light that reaches the plant surfaces, however, is important. Although many injuries attributed to light are probably the result of high temperatures accompanying high light intensities, certain plant injuries have been shown to be caused by light of short wavelengths including the ultraviolet region. The best-documented such diseases is the sunscald of pods of beans grown at high altitudes, where, due to absence of dust, etc., more light of short wavelengths reaches the earth. The pods develop small water-soaked spots which quickly become brown or reddish brown and shrink.

Air Pollution

The air at the earth's surface consists primarily of nitrogen and oxygen (78% and 21% respectively). Much of the remaining 1% is water vapor and carbon dioxide. Man's activities in generating energy, manufacturing goods, and disposing of wastes result in the release into the atmosphere of a number of pollutants which may alter plant metabolism and induce diseases. Air-pollution damage to plants, especially around certain types of factories, has been recognised for almost a century. Its extent and importance, however, increased with the industrial revolution and will apparently, continue to increase with the world's increasing population, industrialisation, and urbanisation. Almost all air pollutants causing plant injury are gases, but some particulate matter or dusts may also affect vegetation.

Some gas contaminants, such as ethylene, ammonia, chlorine, and sometimes mercury vapors, exert their injurious effects over limited areas only. Most frequently they affect plants or plant products stored in poorly ventilated warehouses in which the pollutants are produced by the plants themselves (ethylene), or from leaks in the cooling system (ammonia). More serious and widespread damage is caused to plants in the field by chemicals such as hydrogen fluoride, nitrogen dioxide, ozone, peroxyacyl nitrates, sulfur dioxide and particulates. High concentrations of or long exposure to these chemicals causes visible and sometimes characteristic symptoms (e.g., necrosis) on the affected plants. However, when plants are exposed to dosages less than those the cause acute damage, their growth and productivity may still be suppressed due to interference by the pollutants with the metabolism of the plant. The main pollutants, their sources and their effects on plants are discussed briefly below.

Hydrogen Fluoride

Fluorides are emitted from the stacks of factories and are spread by diffusion or carried by air currents. Hydrogen fluoride (HF) is very toxic to plants, e.g., corn, peach, tulip, on which it can cause injury in concentrations as low as 0.1-0.2 parts per billion (ppb). Fluoride accumulation in the foliage generally injures the leaf margins of dicotyledonous plants and tips of leaves of monocotyledonous plants. Injured ares turn tan to dark brown, die, and may fall from the leaf. Plants differ in the air sensitivity to fluoride.

The more tolerant ones are able to accumulate much more fluoride (up to 200 ppm) without showing necrosis. Instead, they develop a slight chlorosis, followed sometimes by premature defoliation. Actively growing plants, especially when the leaves are wet, are generally the more susceptible to fluoride damage. Fluoride seems to be absorbed by the leaves through the cuticle and be translocated to the leaf margins and tips. When a toxic concentration is reached, the cells from epidermis to epidermis collapse and die. Fluoride may also escape from the plant through volatilization and washing, and the plants then may recover from chronic fluoride symptoms within a few weeks.

Sulfur Dioxide

Sulfur dioxide (SO_2) is also produced from industrial or combustion sources. Sulfur dioxide is itself phytotoxic, but it may also combine with moisture and form acid droplets which, upon settling on plants, can cause injury. The sulfur dioxide gas bleaches the interveinal tissues of leaves of plants such as alfalfa, violet, conifers, etc. The interveinal areas become while to light tan and may later become brown, while the veins themselves remain green. Low concentrations of sulfur dioxide may cause chlorosis of leaves without formation of necrotic lesions. Sulfur dioxide may injure plants in concentrations as low as 0.3-0.5 ppm. Since sulfur dioxide is absorbed through the leaf stomata, conditions that favour or inhibit the opening of stomata similarly affect the amount of sulfur dioxide absorbed. After absorption by the leaf, sulfur dioxide reacts with water and forms phytotoxic sulfite ions. The latter, however, are slowly oxidised in the cell to produce harmless sulfate ions. Thus, if the rate of sulfur dioxide absorption is slow enough, the plant may be able to protect itself from the buildup of phytotoxic sulfites.

Nitrogen Dioxide

Nitrogen dioxide (NO_2) is produced from oxygen and nitrogen in the air by the combustion sources, such as open fires, furnaces, and automobile combustion chambers. Nitrogen dioxide in concentrations of 2-3 ppm causes bleaching of plants similar to that caused by sulfur dioxide. At even smaller concentrations, however, it suppresses the growth of plants such as beans and tomatoes.

Ozone

Ozone (O_3) is one of the most widely occurring air pollutants and apparently one of the most destructive to plants. Ozone originates primarily from the activities of man and his civilization, but it may also be brought to the earth's surface from the ozone-rich stratosphere by vertical winds, it may form from electrical discharges, such as lightning during thunderstorms, and it may be a by-product of photochemical reactions between nitrogen oxides and plant-emanating terpenes, especially in conifer forests. Exhausts of automobiles and other internal combustion engines are probably the most important sources of ozone and other phytotoxic pollutants. Thousands of tons of incompletely burned hydrocarbons and NO_2 are released into the atmosphere daily by automobile exhausts. In the presence of ultraviolet light from the sun, this nitrogen dioxide reacts with oxygen and forms ozone and nitric oxide. The ozone may react with nitirc oxide to form the original compounds:

$$NO_2 + O_2 \xrightarrow{\text{Sunlight}} O_3 + NO$$

However, in the presence of unburned hydrocarbon radicals, the nitric oxide reacts with these instead of ozone, and therefore the ozone concentration bids up. Ozone, too, can react with vapors of certain unsaturated hydrocarbons, but the products of such reactions (various organic peroxides) are also toxic to plants. Normally, the noxious fumes produced by automobiles, and other engines are swept up by the warm air current from the earth's surface rising into the cooler air above, where the fumes are dissipated.

During periods of calm stagnant weather, however, an inversion layer of warm air is formed above the cooler air and this prevents the upward dispersion of atmospheric pollutants. The pollutants then are trapped near the grounds where, after sufficient buildup, they may seriously damage living organisms. Ozone causes a stippling, mottling, and chlorosis of leaves that is confined primarily to the upper leaf surface. The spots may be small or quite large and may vary in colour from bleached white to tan, brown, or black, depending on the plant and on the severity of injury. Many different kinds of plants, including tobacco, alfalfa, bean, cereals, petunia, pine, citrus, have been found to be afflicted with ozone injury in the field. In some plants, such as citrus, grapes, and pine, ozone injury also causes premature defoliation and stunting. Ozone enters leaves through stomata. Once in the leaf, it concentrates in the palisade layer, where it causes collapse and bleaching or discolouration of the palisade cells.

Ozone affects primarily expanding leaves, but not very young or old, mature leaves. Several mechanisms by which ozone can damage plants have been suggested including inhibition of mitochondrial activity, destruction of the permeability of the cell membrane, inactivation of auxin, inhibition of photosynthesis, and inhibition of protein synthesis. Although each of these effects has been observed in at least some of the hosts affected, it is not clear how ozone brings these about. Plants can be protected from ozone damage in several ways. For example, plants escaped damage from ozone when they were watered with ascorbic acid (an inhibitor of oxidation in cells), when sprayed with antioxidants and antiozonants, or when sprayed with dithiocarbamates. Also some plant varieties are considerably more resistant to ozone injury than others.

Peroxyacyl Nitrates (PAN)

Peroxyacyl nitrates nitrates (PAN) cause a plant disorder, generally known as "silver leaf," which produces spots on the lower leaf surfaces of plants of many herbaceous crops. The colour of the spots may range from bleached white to bronze. Peroxyacyl nitrate is produced, along with peroxyacyl nitrate and peroxybutyryl nitrate, the other members of the homologous series of peroxyacyl nitrates, from the reaction of vapors of gasoline or incompletely burned hydrocarbons, produced from the exhaust pipes of cars, with ozone, nitrogen dioxide and probably other oxidizing substances in the presence of sunlight.

The silvering or glazing of the lower surfaces of leaves injured by PAN results from air filling the space created by dehydration and shrinking of the mesophyll cells. Meanwhile the guard cells become congested and the epidermis cells collapse. Although injury is generally limited to the spongy parenchyma of the leaves, the palisade layer may also be affected. In that case, chlorotic symptoms resembling those caused by ozone may appear on the upper surface also. PAN injury has been observed primarily around metropolitan areas where large amounts of hydrocarbons are released into the air from automobiles. The problem is especially serious in areas like Los Angeles and New Jersey, where the atmospheric conditions are conducive to inversion layer formation. Many different kinds of plants, including spinach and petunia, are affected by PAN over large geographical areas surrounding the locus of PAN formation, due to diffusion or to dispersal of the pollutants by light air currents.

Particualte Matter

Plants near roads or cement factories, for example, receive large amounts of dust particles which are deposited on the leaf surface and interfere with the carbon dioxide absorption of the leaves by forming thick crust deposits. Affected plants may become chlorotic, grow poorly, and even die. Additional damage to plants is caused by the toxicity of some of the deposits to leaf tissues either directly or after formation of toxic solutions in the presence of free moisture on the plant.

Nutritional Deficiencies in Plants

Plants require several mineral elements for normal growth. Some elements, such as nitrogen, phosphorous, potassium, calcium, magnesium, and sulfur, needed in relatively large amounts, are called "major" elements, while other, like iron, boron, manganese, zinc, copper, molybdenum, and chlorine, needed in very small amounts, are called "trace" or "minor" elements or "micronutrients." Both major and trace elements are essential to the plant. When they are present in the plant in amounts smaller than the minimum levels required for normal plant growth, the plant becomes diseased and exhibits various external and internal symptoms.

The symptoms may appear on any or all organs of the plant, including leaves, stems, roots, flowers, fruits and seeds. The kinds of symptoms produced by deficiency of a certain nutrient depend primarily on the functions of that particular element in the plant. These functions presumably are inhibited or interfered with when the element is limiting. Certain symptoms are the same in deficiency of an y of several elements, but other diagnostic features usually accompany a deficiency of a particular element. Numerous plant diseases occur annually in most agricultural crops in many locations due to reduced amounts or reduced availability of one or more of the essential elements in the soils where the plants are gown. The presence of lower-than-normal amounts of most essential elements usually results in merely a reduction in growth and yield. When the deficiency is greater than a certain critical level, however, the plants develop acute or chronic symptoms and may even die. Some of the general deficiency symptoms caused by each essential element, the possible functions affected and some examples of common deficiency disorders are given below.

Phosphorus

The plant grows poorly and the leaves are bluish green with purple tints. The lower leaves sometimes turn light bronze with purple or brown spots. The shoots are short and thin, upright and spindly. Phosphorous is associated with almost as many of the same vital functions in the cell as nitrogen. Deficiency in phosphorus interferes with the performance of these functions. Phosphorus is a constituent of nucleic acids, phospholipids, and most proteins and is necessary for the metabolism of carbohydrates, fats, and proteins and for respiration.

Nitrogen

The plant grows poorly and is light green in color. The lower leaves turn yellow or light brown and the stems are short and slender. Nitrogen is essential for proteins (including enzymes), chlorophyll and numerous other plant compounds and, therefore, nitrogen deficiency affects plant growth in many ways at once.

Magnesium

Fist the older leaves and then the younger ones become mottled or chlorotic, followed by reddening and sometimes, appearance of necrotic spots. The tips and margins of leaves may turn upward so that the leaves appear cupped. Defoliation may follow. Magnesium is a structural component of chlorophyll and the cofactor for many enzymes involved in carbohydrate synthesis. Magnesium deficiency, therefore, results in reduced chlorophyll synthesis and chlorosis.

Potassium

The plant has thin shoots. In severe cases dieback may occur. Older leaves may show slight chlorosis with typical browning of the tips, scorching of the margins, and many brown spots usually near the margins. Potassium seems to be essential to many plant functions, including synthesis of carbohydrates and proteins, regulation of cell hydration, and catalysis of reactions, but its exact role is not well-understood.

Sulfur

The plant has pale green or light yellow young leaves without spot formation. These symptoms resemble those of nitrogen deficiency. Sulfur is a components of some amino acids, vitamins, and coenzymes, and seems to be related to chlorophyll formation although it is not a constituent of the chlorophyll molecule.

Calcium

Young leaves become distorted, with the tips hooked back and the margins curled. Often the leaves are irregular in shape and ragged, with brown scorching or spotting. Terminal bids finally die. The plants have poor, bare root systems. Calcium regulates the permeability of membranes, forms salts with pectins in the middle lamella and cell walls, and influences the activity of several enzymes active in the meristematic cells of the growing points. Its deficiency, then, interferes with these functions.

Boron

The bases of young leaves of terminal buds become light green and finally break down. Stems and leaves may become distorted. Fruit or or other large storage tissues may crack on the surface or rot in the center. Many plant diseases, such as heart rot of sugar beets, brown heart of turnips, browning or hollow stem of cauliflower, cracked stem of celery, corky spot of apples, hard fruit of citrus, top sickness of tobacco, etc., are caused by boron deficiency. The role of boron on plants is not known, but it appears to be involved in translocation of sugars, and perhaps in utilization of calcium in cell wall formation.

Iron

The young leaves become severely chlorotic but their main veins remain characteristically green. This effect is probably due to the catalytic role of iron in chlorophyll synthesis. Iron is also a constituent of many enzymes of respiration and other oxidation systems.

Zinc

Leaves show interveinal chlorosis. Late they become necrotic and show purple pigmentation. Few and small leaves, short internodes and low fruit production are common. Zinc deficiency causes "little leaf" of apple, "sickle leaf" of cacao, "white tip" of maize, etc. Zinc is found in plants mostly as a component of several enzymes involved in auxin synthesis and in the oxidation of carbohydrates.

Molybdenum

Leaves develop chlorotic mottling, necrosis, and suppression of the laminate, which may become thin and dry like paper. Growing tips may become distorted and die. These effects are enhanced when nitrogen is supplied as nitrate, because molybdenum is a constituent of the enzyme that reduces nitrate to nitrite.

Manganese

Leaves become chlorotic, but their smallest veins remain green, producing a checked affect. Necrotic spots may appear scattered on the leaf. Severely affected leaves turn brown and wither. Manganese is a cofactor of many enzymes involved in cellular respiration, photosynthesis and nitrogen metabolism.

Copper

The tips of young leaves of cereals wither and their margins become chlorotic. Leaves may fail to unroll and tend to appear wilted. Heading is reduced and the head are dwarfed and distorted. Stone and pome fruit trees show dieback of twigs in the summer, burning of leaf margins, chlorosis, rosetting etc. Copper is a cofactor in many oxidative enzymes.

Soil Minerals Toxic to Plants

Soils often contain excessive amounts of certain essential or non-essential elements, both of which at high concentration may be injurious to the plant. Of the essential elements, those required by plant in large amounts, such as nitrogen, phosphorus, potassium, etc., are usually much less toxic when present in excess than are the elements required only trace amounts, such as manganese, molybdenum, boron. Even among the latter, however, some trace elements such as manganese and molybdenum have a much wider range of safety than do others e.g., boron. Besides, not only do the elements differ their ranges of toxicitry but various kinds of plants also differ in their susceptibility to the toxicity to a certain level of a particular element.

Concentrations at which non-essential elements are toxic also vary among elements, and plants in turn vary in their sensitivity to them. For example, some plants are injured by very small amounts of nickel, cobalt, or chromium, but can tolerate considerable concentration of aluminium or selenium. The injury occurring from excess of an element may be slight or severe. It may be the result of direct injury by the element to the protoplast of the plant cells, including interference with vital enzyme or other basic functions of the cell. One the other hand, the elements interfere with the absorption or function of another element and thereby lead to the symptoms of a deficiency of the element being interfered with.

Thus, excessive sodium induces a deficiency of calcium in the plant, while the toxicity of copper, manganese, zinc, chromium, cobalt, or nickel is both direct on the plant and by inducing a deficiency of iron in the plant. Excessive amounts of sodium slats, especially sodium chloride, sodium sulfate, and sodium carbonate, raise the pH of the soil and cause what is known as alkali injury. This injury varies in the different plants and may range from chlorosis to stunting, leaf burn, wilting, to outright killing of seedlings and young plants. Some plants, e.g., wheat, apple, are very sensitive to alkali injury, while others e.g., sugar beets, alfalfa, and several grasses, are quite tolerant.

On the other hand, when the soils is too acidic, the growth of some kinds of plants is impaired and various symptoms may appear. Plants usually grow well in a soil pH range from 4 to 8, but some plants grow better on the lower pH than others, and vice versa. Thus, blueberries grow well on acid soils, while alfalfa grows best on alkaline soils. The injury caused by low pH is not always the result of high hydrogen-ion concentration; in most cases, it is brought about the greater solubility of mineral salts in acid solutions. These salts then become available in concentrations that, as was pointed out above, either are toxic to the plants or interfere with the absorption of other necessary elements and so cause symptoms of mineral deficiency. Boron, manganese, and copper have been most frequently implicated in internal toxicity diseases, although other minerals, e.g., aluminum an iron, also damage plants in acid soils. Excess boron is toxic to many vegetables and trees.

Excess manganese is known to cause a crinkle-leaf disease in cotton, and has been implicated in the internal bark-necrosis of Red Delicious apple and in many other diseases of several crop plans. Sodium and chlorine ions also have been shown to cause symptoms of poor growth and decline, like those shown by some of the trees along roads in northern areas where heavy salting is carried out in the winter to remove ice from roads.

Improper Agricultural Practices

A variety of agricultural practices improperly carried out may cause considerable damage to plans and increased financial losses. Almost every agricultural practice can cause damage when done the wrong way, at the wrong time, or with the wrong materials. Most commonly, however, losses result from application of chemicals, such as fungicides, insecticides, herbicides, fertilizer, at too high concentrations or on plants sensitive to them. Spray injury resulting in leaf burn or spotting or russeting of fruit is common on many crop plants.

6

Spreading Diseases

Epidemics

This chapter is analytical rather than descriptive. This chapter aims at an understanding of the factors and processes that go to make an epidemic, and not at a description of important epidemics, past or present.

A. Some Common Misconceptions

It is stated in the literature that for an epidemic to occur there must be an aggressive parasite that multiplies fast, spreads for and swiftly, and is not particularly selective in its requirements. These are total misconceptions. They are the cause of much confused thinking and must be disposed of forthwith. Consider swollen shoot of cacao, a systemic virus disease in West Africa, as an example. The virus complex is carried by slow-moving flightless mealy bugs of the family Pseudococcidae, and spreads largely from tree to tree in contact. It is delicate, and survives for less than an hour in the feeding vectors. The number of infected trees on a farm multiplies slowly; under conditions of unrestricted natural spread it took 30 months for the percentage of infected trees to increase from 31 to 75.

The spread of infection from farm to farm is also slow. In 1947 the largest area of disease in West Africa had reached a radius of only some 10 miles after having spread continuously since 1922. Yet there can be no doubt that swollen shoot is a major epidemic. The virus complex is endemic in West Africa indigenous tree of the Sterculiaceae and Bombacaceae. As a sporadic disease, cacao swollen shoot has a long history in West African. But some 30 years ago the epidemic began, with the merging of separate but expanding, more or less circular outbreaks, into larger amorphous areas of dead and dying trees, which spread ever more rapidly and devastated whole farms in their spread. The disease has caused political upheavals, and the cacao industry from the western Ivory Coast to central Nigeria has needed drastic measures to save it. Another example is oak wilt, caused by the fungus *Ceratocystis fagacearum*.

The current epidemic of the disease in the United States was rated sufficiently high to get an entire chapter to itself in the limited space of "Plant Diseases," the United States Department of Agriculture's Yearbook for 1953. Yet the parasite has very limited means of spread. It can spread over a distance by some means that are not yet properly understood, but the main spread is from tree to tree. Infection takes place through root grafts, and local barriers such as roads are sometimes enough to stop it. An extreme example is psorosis, a virus disease of citrus. No vector is known. The virus can spread from tree to tree by root grafts, but the natural spread is so slow that it is extremely difficult to demonstrate in orchards. Yet psorosis is currently the greatest killer of citrus trees in California. Psorosis epidemics, it seems, may take half a century or more to develop.

B. Epidemics as a Matter of Balance between Opposing Processes

The human population of a country can increase as a result of a high birth rate; it can also increase as the result of a low death rate. It will increase with a low birth rate provided that the death rate is even lower. As with men, so it is with plant pathogens. Disease can increase to epidemic levels because (to consider only the extremes) the pathogen has a high birth rate or because it has a low death rate. The misconception we have just discussed have arisen because attention has been focused on high birth rate epidemics. But low death rate epidemics are also important, particularly with perennial hosts. High birth epidemics are usually caused by fungi which develop uredospores, conidia, oidia, and the like, and produce local lesions in the host.

Low death rate epidemics are caused largely but not entirely by systemic pathogens. The explanation is often fairly obvious. The systemic pathogen is safe within the host, and if the host is perennial and the disease not immediately lethal, a long life of the infected host guarantees a long life to the systemic pathogen. With pathogens as well as with men longevity means a low death rate. The terms are practically equivalent. The line between the two types of epidemic is often somewhat blurred, but at the extremes the distinction between them is clear and has interesting consequences. High birth rate epidemics are usually controlled by fungicides or resistant varieties; low death rate epidemics are mainly controlled by means of sanitation. High birth rate epidemics usually spread fast; low death rate epidemics usually spread slowly. These and other consequences will unfold themselves logically as we proceed. But the terms high birth rate epidemics and low death rate epidemics have been used purely to illustrate a point and now will be dropped in favor of more conventional expressions

I. Multiplication of Infections

A. Multiplication within a Crop

1. *The Course of an Epidemic*. There is no such thing as a typical epidemic. The variety of epidemics in plant pathogen is infinite. To mention just two factors, the epidemic varies with the amount of inoculum that is the source of the infection, and this ranges with the different diseases from scarce to abundant; it varies with the rate at which the infections multiply, and this ranges from very slow to very fast. Because of this variety, one example is as good as another. Blight of potatoes, caused by *Phytophthora infestans*, is chosen here as an example because of its place in the history of plant pathology. But even for this on disease the example we have chosen for illustration is not typical of all epidemics. It is based on data from western Europe.

If blight in North America had been the example, stress would have been laid on the danger of epidemics starting from piles of culled potatoes and garbage dumps near towns. Fig. 6.1. is based on data taken from a probit foliage decay line given by Large (1945) for an epidemic of blight on potatoes in England. Observations were started on august 11, when 0.1% of the foliage was infected. This is equivalent to an average of about 1 lesion per plant. The epidemic progressed for about 4 weeks, after which all the foliage had been destroyed by blight. The curve that shows this in Figure is sigmoid. The figure also shows the rate of increase of infection per cent per day. The rate for this epidemic began at about 48%, when observations were started on August 11, and gradually dropped to zero as the epidemic ran its course. In the Netherlands the early history of a potato blight epidemic was studied by Van der Zaag (1956).

We assume that the results hold for England too Van der Zaag found that the most important primary foci of infection were infected plants growing from infected seed. These plants had a few weak, shriveled shoots with a few lesions from which spores were released during the second half of May. They were most abundant in the every

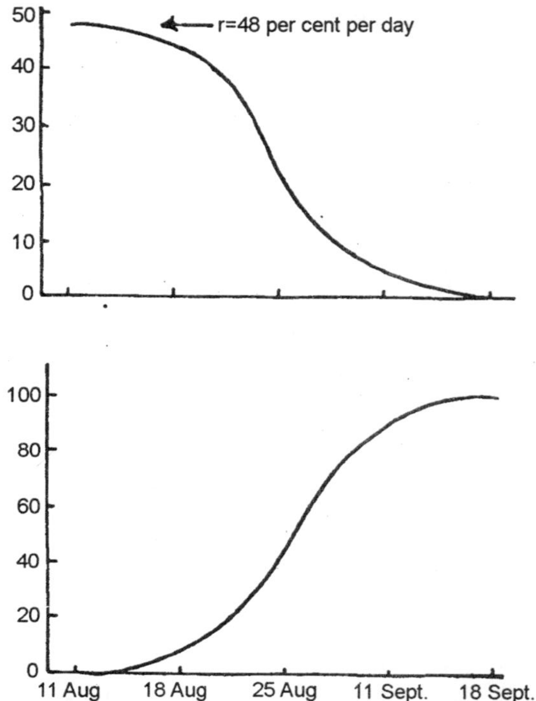

Fig. 6.1. Lower half. The progress of an epidemic of blight (P. infestans) in potatoes. Upper half. The increase of infection per day expressed as a percentage of the infection already present.

susceptible variety, Duke of York, in which about 1 primary focus per square kilometer was found. Infection spread first around these foci, then generally over the fields, and from field to field and from variety to variety. Our concern here is not with the progress in any particular field but with the epidemic generally.

There are 3,000,000 or 4,000,000 plants in 1 sq. km. of potatoes, so that there are (in round figures) 3,000,000 or 4,000,000 lesions per square kilometer of potatoes when 0.1% of the foliage is infected. Still in round figures, infection had to multiply 1,000,000 times from the original foci before the level of 0.1% was reached on August 11. The one-thousandfold increase (from 0.1 to 100%) that occurred after August 11, and ignores the one-millionfold increase that went before. And in terms of time starts in August instead of May, like the biography of a centenarian that starts after his seventieth birthday. Consider the matter from the farmer's point of view. For him the most important single characteristic of an epidemic is its date of onset. This decides when he must spray or whether he need spray at all. If the date is early—say, in July in England—heavy losses are likely to occur if the disease is unchecked. But if it is late—say, in September—losses are likely to be small (apart from tuber infection), and spraying would be a waste of material and effort.

There are of course some variations in detail (an epidemic may start early and then be checked by a change in the weather) but the general pattern is clear, as can be seen from the analyses of Moore (1943), Large (1952), Beaumont et. al. (1953), and Large et. al. (1954). The date of onset is determined by what happens before the onset. Infection increased one-millionfold between the second half of May and August 11, i.e., in about 80 days. This is equivalent to an average increase "at compound interest" at the rate of 17% per day. But the rate would not have remained at a steady average since it is affected by changes in the weather. Apart from that, the potato becomes more susceptible as it becomes older. One can therefore infer that the rate was considerably less than 17% at the start and increased to 48% by August 11.

2. *Rate of Multiplication before the Onset of the Epidemic*. For convenience we can define the onset of an epidemic as that time when not more than 5% of the susceptible tissue is infected, or, for systemic diseases, when not more than 5% of the plants are infected. Within this rough limit we can decide on any criterion we wish; for example, with potato in England and Wales the starting point for the assessment of blight is 0.1% infection, and this level can conveniently be taken to indicate the onset. The rate of increase of disease has been taken as

$$\frac{dx}{dt} = kx\,(1 - x) \qquad (1)$$

in which k is a constant and x is the proportion of susceptible tissue that is infected, or, for systemic diseases, the proportion of infected plants. The rate at which infection occurs is, according to this equation, proportional to the proportion of infected tissue and the proportion of healthy susceptible tissue, that is, to the proportion of tissue that is still available for infection. Curves of this equation describing the progress of infection with time have the sigmoid shape so commonly found in plant pathology. Large (1945) applied the equation to blight epidemics and concluded that it would give a good fit to actual blight progress curves within the limits of accuracy of the original blight observations. Nevertheless the reader is warned that after the onset a good fit can come about only as a result of counterbalancing factors, because the incubation period (i.e., the pre-reproduction period after infection) inevitably tends to increase the value of k as the epidemic progresses beyond the onset. That is, after the onset, k is just an empirical constant even in a constant environment. We use equation 1 and sigmoid progress curves almost solely to provide a helpful introduction to the study of multiplication. After the introduction the topic is treated in a way that will nearly obviate the need for considering disease progress curves beyond the onset.

Infection is influenced by the weather, the susceptibility of the host, the virulence of the pathogen, etc. So equation 1 will be rewritten in the form

$$\frac{dx}{dt} = kx\,(1-x)\,f(T, H, S...) \tag{2}$$

in which $f\,(T,\ H,\ S\ ...)$ in a function of the temperature and humidity of the air, the susceptibility of the host, and so on.

Before the onset of an epidemic, x is small and, in the definition already given for the onset, less than 0.05. As a good approximation for the progress of infection up to the onset, the term $(1 - x)$ can be dropped and equation 2 rewritten as

$$\frac{dx}{dt} = kxf\,(T, H, S\,...) \tag{3}$$

$$r = \frac{100}{x}\frac{dx}{dt}$$
$$= 100kf\,(T,\ H,\ S\ ...) \tag{4}$$

By definition, r is the rate of increase per cent per day (or per year, if years are the units of time) up to the onset of the epidemic. Within the limits of the approximation just stated r is independent of the progress of the epidemic (up to the onset) and becomes a direct reflection of the environmental conditions, the susceptibility of the host, and so on. A qualification, normally of minor importance.

3. *Some Comments about r.* One must become familiar with r. It is a fundamental and perhaps the most useful concept of epidemiology. It plays its part not only in multiplication rates, but also in the rate of spread of epidemics in distance. (Section III, A and E), in the theory of forecasting (Section II, B), in sanitation (Section II, C), and directly or indirectly, in much else. The impression may have been given that r is only an approximation. That is incorrect. For convenience r is allowed to make its debut approximately as the percentage rate of multiplication at or before the onset of an epidemic, which is accurate

enough for present purposes. But r is a concept capable of exact determination (as by equation 5) and applicable throughout the course of the epidemic. The reasons for using it mainly at or before the onset will appear later.

In reporting the results of censuses, human birth rates, death rates, and rates of increase of population are given in terms of 1,000 of population. It is not assumed that the rates are the same for all hours of the day, for all seasons of the year, or for all years of boom and depression, peace and war. The crude rates are simply a convenient way of expressing what is happening. So too r can be regarded as a convenient way of reporting the results of censuses of infections before or at the onset. It is not implied that r is constant from hour to hour or day to day. It just expresses the tempo of developments to the onset in an uncomplicated way. But one condition is implied: fractions of a day or, if appropriate, a year must be avoided.

When, for example, it is calculated in Section II, C that an epidemic would be delayed 3.7 days this must be interpreted as "from 3 to 4 days." In one way r has an advantage over crude census rtes, which are subject to the influence of varying proportions of the population in the different age groups, particularly to a varying proportion of women of childbearing age. Superficially, the same problem occurs with r. With potato blight, for example, it takes from 4 to 6 days or longer after inoculation for spores to be produced. But for most of this period there is no recognizable lesion, and spore formation begins soon after the lesions become visible in the field. Up to the time of the onset of the epidemic relatively few of the lesions seen in the field are not of spore bearing age; one could without great difficulty count as lesions only those of an age to produce spores.

The validity of using lesions of this sort for determining multiplication rates is examined in Section II, A, 6. The magnitude of r reflects the effect of temperature, humidity, rainfall, wind, and other environmental factors on multiplication; it reflects among other factors the resistance or susceptibility of the host, the proportion of spores that germinate, the proportion of germinated spores that manage to enter the host and establish infection, the rate of increase in size of the lesion, the speed at which spores are produced and their number, the abundance of vectors and their mobility, efficiency, and distribution, and the size and number per acre of the host plants.

It summarizes the equilibrium of the factors (except x itself) that affect the tempo of multiplication. Because of this, few of these factors will be specifically dealt with in this chapter. There is, for example, no discussion of the effect of weather and climate. This may seem strange in a chapter on epidemiology. But it is logical. When a forecaster uses his knowledge of the weather and other relevant factors to forecast an epidemic, he is in fact forecasting a faster rate of multiplication. To him the weather and other factors are important. To us (for the purpose of this chapter) the rate of multiplication is important. This statement carries no implication that the one approaches to the discussion of epidemics. One chooses what one believes to be the most apt for one's purpose.

4. *The Estimation of r.* Substitute $r/100$ for $kf\ (T, H, S\ ...)$ in equation 2, integrate, change logarithms to the base 10, and rearrange thus:

$$r = \frac{230}{t_2 - t_1} \log \frac{x_2\,(1 - x_1)}{x_1\,(1 - x_2)} \qquad (5)$$

Here x_1 and x_2 are the proportion of susceptible tissue infected (or, for systemic diseases, the proportion of infected plants) at dates t_1 and t_2, and $t_2 - t_1$ is the interval in days (or years, if this is appropriate) between these dates.

For example, according to the data of Large (1945) blight in potatoes at Dartington in 1942 increased from 0.1 to 5.0% in 7 days. Here x_1 and x_2 are 0.001 and 0.05, respectively, and $t_2 - t_1$ is 7 days. From equation 5, r = 57% per day. In general, the use of equation 5 assumes that the lesions are randomly distributed. In the particular case of potato blight there is the independent evidence of Gregory (1948) that this is so; but randomness is not the rule. There are often quick departures from randomness, especially when systemic diseases multiply, particularly the systemic diseases of trees. There is a general rule about this: the smaller the size of lesion, the higher the value of x_2 that can be used without incurring a serious error from lack of randomness. A systemically infected plant must be considered as a single lesion (Section III, F).

As another example, consider cauliflower mosaic. It spreads in nonrandom fashion to form nests of infected plants. In a trial with different strains of cauliflower there were 1.7 and 5.9% infected plants on August 21 and September 4, respectively, so x_1 and x_2 are 0.017 and 0.59, respectively, and $t_2 - t_1$ is 14 days; whence r is 9.2% per day. The epidemic was started artificially by planting an infected plant at the center of each plot of 121 plants, so 5.9% infection represents roughly 7 plants around each infector plant. The evidence is that at this stage the error from nonrandonmess is small, so the estimate of r is fairly trustworthy; but at much higher values of x_2 estimates would be too low.

With local lesion disease on plants that grow between t_1 and t_2, allowance must be made for new susceptible tissue. If susceptible tissue increases m times between t_1 and t_2 equation 5 must be rewritten as

$$r = \frac{230}{t_2 - t_1} \log \frac{mx_2 (1 - x_1)}{x_1 (1 - x_2)} \tag{6}$$

Chester (1943) gives figures for the multiplication of leaf rust (*Puccinia triticina*) of wheat in Oklahoma. After a severe winter followed by normal weather, infection increases at a steady rate from 1 pustule per 1,000 leaves at the beginning of March to 10 pustules per leaf at the beginning of June. During this period there is a tenfold increase of leaf tissue; hence m is 10. The ratio x_2 / x_1 is 10,000, and $t_2 - t_1$ is 92 days. An infection of 10 pustules per leaf is equivalent to about 1% infection, so x_2 is 0.01 and x_1 is even smaller, and we can ignore the terms $(1 - x_1)$ and $(1 - x_2)$. By equation 6, $r = 12.5\%$ per day.

5. *General comments on the Restriction of Much of This Chapter to Events before the Onset.* Physical chemistry advanced far in the theory of dilute solutions. The gas laws were applied to dilute solutions, and dissociation constants were determined in dilute solutions. This was done in the first place to avoid the difficulties of dealing in theory and practice with the behavior of molecules and ions at high concentrations. In the same way this chapter deals to a great extent with the theory of dilute concentrations of disease: of disease up to the onset, defined arbitrarily as the 5% level. Primarily the reason for this is to avoid the objection against using disease progress curves after the onset. How much information do we lose by confining quantitative discussions to the period up to the onset? The answer is: surprisingly little. Consider these examples. In Section II C an equation is given that evaluates sanitation in terms of a delay in the onset of the epidemic. If the weather and other conditions stay constant, the delay in the onset is also the delay in reaching the 50% level of disease or the 99% level. If the weather changes, then interest is changed from the factor of sanitation to the factor of weather.

By describing the effect of sanitation on the onset, one has in effect described all that needs to be described about the effect of sanitation on the course of the epidemic. In other words, the loss of information from confining attention to the period up to the onset is, negligible. Also, it does not matter what criterion one takes for the onset, provided that it is not more than about 5%; it could just as well or better be taken at 0.1%. In Section IIB we discuss the application of r to the theory of forecasting; in most cases forecasts are in practice confined to the onset of an epidemic. Gradients of infection away from the source figure largely in Sections III and V. With gradients, too, it can be shown both that it is wise to keep calculations to the period up to the onset of the epidemic and that remarkably little information is lost by doing so. In this chapter we do in fact discuss gradients at higher levels of disease, and correct the data as best we can. But this is only through lack of choice; the data are so scant that one cannot at present afford to be selective.

6. *Justification for Using the Law of Compound Interest when there is an Incubation Period.* The meaning given to r is that it is the rate of compound interest (per cent) up to the onset. True compound interest (we are not concerned with bankers' compound interest) is added to accumulated capital as it is earned, and then instantly begins earning itself; the concept of an incubation (i.e., pre-reproduction) period is foreign to the law of compound interest as it is used in science. Are we then justified in applying the law, as represented by r to plant disease? It seems that we are, and it will be shown that r can be used as an apparent rate. The meaning of x is the proportion of infected tissue in which incubation is complete. The infection is visible and the lesions are of an age to produce inoculum (according to the suggestion in Section I, A, 3). Consider now the total proportion x' of tissue that is infected, even if incubation is still incomplete.

To simplify the argument, we shall confine attention to events before the onset of the epidemic. Instead of equations 3 and 4, let us write for the same set of observations.

$$\frac{dx'}{dt} = \frac{r'}{100} x'_{t-P} \tag{7}$$

Here x'_{t-P} is the proportion of infected tissue at time $t - P$, in which P is the incubation period. It is the proportion of tissue in which incubation is complete and there fore has the same meaning as x in the earlier equations. It can be determined that

$$r = r' e^{-Pr/100} \tag{8}$$

In effect, r is a compound interest rate to which P contributes, outwardly like any other factor, such as the abundance of vectors.

Equation 8 holds only when P and r' are constant. More generally, it can be shown that despite an incubation period the concept of compound interest is valid even if P and r' vary with temperature, humidity, etc. There is more to it than just that. In the early history of an epidemic (for example, soon after inoculation in an artificial epidemic) r at first varies with time, even if P and r' are constant, and only later settles down to the stable value given by equation 8. The larger the product Pr, the larger are the variations.

Similar variations occur when, for example, an epidemic is checked by drought and then builds up again. The previous history of the epidemic, every previous fluctuation, is remembered in the multiplication rate, and this memory factor is the special contribution of an incubation period to the concept of compound interest. There is a qualification. Previous memory is wiped out each time multiplication stops for a continuous interval at least as long as P, and then starts anew. Hence, the memory factor can cause a drift in the multiplication rate of potato blight from day to day, but not in that of potato leaf roll from year to year.

Variations in the value of r, when caused by memory or any other factor, are properly cared for by equation 5. This equation estimates the average of the value of r at every instant between any times t_1 and t_2, irrespective of any variations that occur. But although it is important to know that the memory factor's effect or r is correctly estimated, it is often equally important to know how to eliminate this effect so that other factors can be studied without interference. Except within approximately 2.2 P days after then end of a major interruption in a natural epidemic, the effect of the memory factor on r can be made small by arranging that the interval between t_1 and t_2 should be as near to 1.2 P days (or a multiple thereof) as is possible without using fractions of a day. For example, if P is 6 days and it is wished to study in an artificial epidemic the separate effect on r of factors other than the memory factor, observation should begin whenever convenient after the thirteenth day from inoculation and then be made at weekly intervals. Very exact knowledge of P is here unnecessary, and in the example just quoted an 8-day interval between t_1 and t_2 would also eliminate the memory factor's effect on r.

B. Increase of Disease when the Pathogen Does Not Spread between the Host Plants

Consider bunt of wheat caused by *Tilletia caries* and *T. foetida*. Plants are infected as young seedlings, and do not release spores until the grain has been formed. One plant cannot infect another during the course of the season. There is multiplication of infection but only as an increase from season to season and not within a single crop. Many other systemic smut fungi behave similarly. With obligate heteroecism there are similar restrictions. The apple rust fungi *Gymnosporangium* spp. move hither and thither from apples to the alternate hosts and back. During this movement from one host tot he other there can be a multiplication of infection, but there is no multiplication directly from apple to apple. Similarly, *Cronartium ribicola* does not spread from pine to pine.

With many of the systemic or quasi systemic vascular wilt diseases the pathogen is not returned in great quantity to the soil until the host dies. The pathogen may build up from season to season, but it does not spread much from plant to plant during the season. Sometimes there is apparently no spread from plant to plant and no building-up within the host species even over the seasons. Eastern X-virus spreads form chokecherry (*Prunus virginiana*) to peach but apparently not from peach to peach. The virus of Pierce's disease has a wide range of species of host plants from which it infects grapevines, but it does not seem to spread from grapevine to grapevine.

Tomato spotted wilt virus infects tobacco, but the thrips vectors do not breed on this host and there is no evidence of spread from tobacco to tobacco. Peaches with X disease, grapevines with Pierce's disease, and tobacco with tomato spotted wilt apparently do not contribute to the building-up of an epidemic of these three virus diseases; in them epidemics are secondary and are the result of multiplication in other hosts. Absence of spread from plant to plant can occur at times even with diseases that normally spread. Figure 6.2A, reproduced from a report by Doncaster and Gregory (1948), shows the distribution of rugose mosaic, caused by virus Y, in an isolated potato field. This field was initially free from virus Y, but was invaded from a source 300 yd. away. The plants which were infected as a result of this invasion did not pass the infection on to their neighbors, and there was no evidence of secondary spread. Absence of secondary spread can be expected with the invasion takes place late in the season or when the presence of vectors is transient. It is possibly not uncommon. The type of increase without multiplication which has been discussed in the previous five paragraphs can cause an epidemic which superficially resembles other epidemics but which is fundamentally distinct. There is, for example, no reason why the progress curve of the epidemic with time should be sigmoid. Often the approach to control is different.

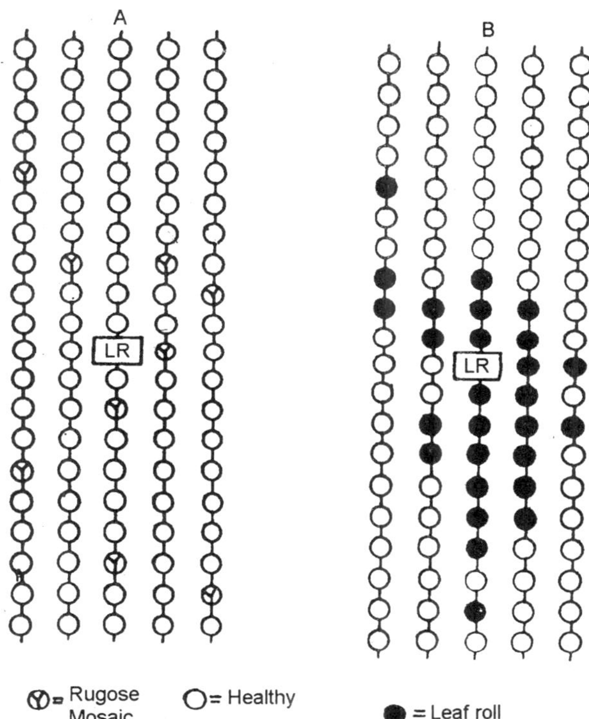

Fig. 6.2. Distribution of rugose mosaic and leaf roll around and infected plant. L.R., grown from a potato tuber infected with leaf roll. For convenience, rugose mosaic and leaf roll are shown in separate diagrams, A and B, respectively.

It is therefore important to be able to detect when infection is increasing without spreading from plant to plant. This can be done in various ways. One can mark infected plants and determine whether they are foci of infection by comparing the number of plants which become infected in an area near them with the number of plants which become infected in a comparable area at a distance away from them. But the most convenient methods use the distribution of diseased plants in the field or orchard without a knowledge of their history. Figure 9.2B shows how leaf roll develops in a potato field around an infected plant grown from an infected tuber. There is a nest of diseased plants near this plant. Nests of this sort vary in their compactness and shape from disease to disease, but it is probably safe to assume that with all plant disease there is a tendency for infected plants to aggregate to some extent near the source of infection. (The tendency is greatest when r' is low in value and the gradients are steep; see Section III E and F about scales of distance). The test for spread of infection is a test for aggregations. Cochran (1936) examined some general tests, and Todd (1940) and Freeman (1953) developed a test for the special case of the death of trees planted at the corners of a square lattice in a rectangular plantation. The simplest test yet devised makes use of doublets. A doublet consists of 2 adjacent diseased plants. If n plants are examined in sequence, and of these μ are diseased, the expected number of doublets is

$$d = \frac{1}{n}\mu(\mu - 1) \tag{9}$$

A run of 3 diseased plants is counted as 2 doublets; a run of 4, as 3 doublets; and so on. At test of significance has been worked out by G.A. McIntyre of the Commonwealth Scientific and Industrial Research Organisation, Canberra, Australia, and it is hoped it will be published shortly. As an example, the distribution of leaf-roll-infected potato plants in Fig. 9.2B is analyzed. There is a total of 100 plants, including the original diseased plant; of these, 28 are infected. If one starts at the top left corner, reads down the first row, up the next, and so on, one counts 19 doublets. The expected number is only 28 × 27/100 or 7.56. The infected plants show excessive (nonrandom) aggregation.

A corresponding analysis for the 9 potato plants with rugose mosaic in Fig. 9.2A gives d (observed) 0 and d (expected) 0.72. There is no evidence of nonrandom aggregation. In practice one is usually concerned with collecting evidence for the absence rather than for the presence of spread of infection. Extensive data are needed, and precautions must be taken against extraneous heterogeneity of the field. The method can be used to examine evidence for spread in any direction, and counting can start from any randomly chosen starting point. Missing plants, uneven spacings, or changing directions do not affect the analysis mathematically, although they may affect its interpretation biologically. From what is written in Section III F it follows that spread of infection is most easily detected by this method if the disease is a systemic disease of trees.

II. The Amount of Inoculum at its Source in Relation to Epidemics

A. The Amount of Inoculum and Rate of Multiplication before the Onset

In the preceding section our concern was with rates of multiplication. The amount of inoculum at the source of the epidemic must now be considered. When infected plants are the starting inoculum—when, for example, epidemics of potato blight start from lesions on infected shoots developing from infected tubers or when potato leaf roll spreads from an infected plant, the sequence of multiplication is homogenous. Lesions produce lesions, and infected plants produce infected plants. We shall be concerned only with homogeneous sequences. When the source of inoculum is resting spores or anything but infected plants, we consider the sequence as starting from the first plants to be infected directly from this inoculum, i.e., the first infected plants are considered as the inoculum at its source. This restriction to homogeneous sequences simplifies what has to be discussed without affecting the conclusions appreciably.

Suppose I_0 and I are respectively the amount of inoculum (in terms of the number of lesions or, for systemic diseases, the number of infected plants) at the source and at the onset of the epidemic which was defined in Section I, A, 2.

$$I = I_0 e^{rt/100} \qquad (10)$$

Here t is the time taken up to the onset, and r has the meaning given in earlier equations. To quote an example given previously, if *P. infestans* multiplies a millionfold in 80 days.

$$e^{80r/100} = I / I_0 = 10^6$$

whence r = 17.3% per day. This is an average rate for the 80 days.

More often one needs to know the effect of a change in the amount of inoculum at the source: a change brought about, for example, by sanitation. One can therefore profitably rewrite equation 10. Suppose the amount of inoculum at the source is reduced from I_0 to I'_0 . The delay in the onset of the epidemic, Δt, can then be determined from the equation

$$I_0 = I'_0 \, e^{r \, \Delta t/100}$$

whence

$$\Delta t = \frac{230}{r} \log \frac{I_0}{I'_0} \qquad (11)$$

In the equation, r is the rate during the time of the delay. Logarithms are to the base 10.

Suppose r= 69.3% per day at the onset. How long will the onset be delayed by having the amount of infection is doubled every day, having the amount of inoculum at its source?

$$t = \frac{230 \log 2}{69.3}$$

$$= \frac{230 \times 0.301}{69.3}$$

$$= 1.0$$

The answer is 1 day. Natural compound interest of 69.3% per day, which is added to the accumulated capital at every instant, is equal to 100% "banker's" compound interest, which is added once a day. This numerical problem makes the obvious point that if the amount at its source will postpone the onset by a day. The criterion for the onset must of course be fixed, but, provided that it is fixed within the limit stated in Section II, A, 2, the particular value at which it is fixed has no relevance. (But read also Section II, A, 5).

Consider another example. Broadbent (1957) tested the use of barriers to protect cauliflower seedlings from mosaic, and found that a barrier three rows of barley around the seedbed reduced infection from 3.0 to 0.6%. How long would be onset of an epidemic of cauliflower mosaic be delayed in crops planted from protected seedbeds? Use the value r = 9.2% per day, calculated for cauliflower mosaic in Section II, A, 5.

$$\Delta t = \frac{230}{9.2} \log \frac{3.0}{0.6}$$

$$= 17.5$$

The onset of the epidemic would be delayed 17.5 days. For properly apt results, r should be measured for the same variety as that used in the barrier experiment, in the same district, at the same time of the year, in the same type of soil, etc.

B. Some Comments on the Theory of Forecasting of Epidemics

The method and effectiveness of forecasting epidemics varies with the value of r in the equation 11. Consider the two extremes of forecasts that ignore, and forecasts based entirely on, the amount of inoculum at the source.

1. *Forecasts That Ignore the Amount of Inoculum.* According to equation 11, the higher the value of r at the onset, the less the error of a forecast of the time of onset that ignores differences in the amount of inoculum at the source. If r is large enough, approximate forecasts become possible without reference to inoculum. An alternative condition

would be that the amount of inoculum is so large that it ceases to be a limiting factor. Both conditions possibly exist with, say, apple scab when scabby spring weather follows a fall and a winter favorable to the survival of large amounts of inoculum in dead leaves. Potato plants become more susceptible to blight as they grow older, and high values of r are possible toward maturity. It is this fact of a high value toward maturity that makes short-range forecasts of blight feasible on weather data alone. It has been found that there may be a "zero time", a date (July 1 in west Scotland) before which forecasts based on weather data alone are invalid (Grainger, 1950). Interpreted by equation 11, "zero time" is the date on which the magnitude of r becomes high enough to blur the effect of variations in the amount of inoculum and make the epidemic specially sensitive to the weather. The more susceptible the variety, the earlier can "zero time" be fixed or, alternatively, if "zero time" is unchanged, the more accurately can forecasts be made.

2. *Forecasts Based on the Amount of Inoculum Alone*. As the value of r at the onset decreases, the effect of variations in the amount of inoculum on the date of the onset increases. With low enough values of r the importance of the inoculum factor becomes dominant, and approximate forecasts can be based on it alone. The number of sclerotia of *Sclerotium rolfsii* in the soil can be used to forecast losses of sugar beet from this fungus. In general the soil is not a medium conducive to the fast multiplication of pathogens, and methods of forecasting other diseases caused by soil-borne pathogens on inoculum alone are likely to be feasible.

C. An Epidemiological View of the Problem of Sanitation: the First Rule of Sanitation

Equation 11 tells us that, measured as a delay in the onset of an epidemic, the benefit of a given percentage reduction in the amount of inoculum at its source is inversely proportional to r, the magnitude of r being determined at the time of the delay. This is the first rule of sanitation. Within the limits of one disease the rule means that sanitation helps most when it is needed least. Before any method of sanitation is recommended for the control of a disease, it should be ascertained that the method will remain effective even during seasons when conditions are such that infection multiplies at its fastest.

In comparing different diseases the rule means that control by sanitation is most apt for those disease that multiply most slowly. If r is large, sanitation is relatively ineffective, and it is usually necessary to use fungicides or, alternatively, to call in the plant breeder to produce a resistant variety, i.e., to bring down the magnitude of *r*. When *r* is small, one is likely to be dealing with a disease that can be controlled by sanitation: by crop rotation, by fumigation of the soil before the crop is planted, by destroying diseased crop residues or using manure free from inoculum, by planting healthy or disinfected seed and nursery stock, by roguing out diseased plants, by isolating fields from sources of inoculum, by destroying weeds or other hosts that can carry infection, or by any other method that reduces inoculum at its source.

For example, systemic diseases tend to have low values of *r*, and a study of the literature shows that the majority of them are controlled by sanitation in one form or other. As *r* increases, sanitation becomes less effective. But where do we draw the line? At the one extreme, with very low rates of multiplication, sanitation is almost completely effective; for example, on present knowledge, a citrus grower can be protected against losses from psorosis by the sanitary measure of using only healthy nursery stock to plant his orchards. Are there, at the other extreme, infections that multiply so fast that sanitation is worthless? The answer depends on the circumstances. Consider this example.

A field of potatoes grown from blight free seed can become infected with blight from infected refuse piles or from infected fields. Take the second alternative. From calculations based on meager data in the literature, it can be estimated that doubling the isolation of a potato field from its neighbors would reduce the amount of blight that the field gets from its neighbors by 83%. (See Section III B and D). If we take *r* to be 48% per day, this isolation will postpone the onset of an epidemic by 3.7 days. (Here $I_0/I'_0 = 100/(100 - 83)$). On an allowance of an increase of a ton of tubers per acre per week during the critical period of the growing season, isolation to the extent stated would mean a gain of crop in an unsprayed field of about half a ton per acre. Whether this gain would be worth working for depends on the circumstances. On a small farm, isolation is impossible or possible only at the cost of arrangements that are not acceptable to the farmer. But on a large estate some farm planning with an eye to isolation is not ruled out. It is not for our present purpose desirable to pursue this topic at length, but it should be stated that even with a fast multiplying disease like blight of potatoes the possibility of sanitation cannot be excluded before conditions are analyzed.

D. An Epidemiological View of Breeding for Resistance

One especially important deduction from equation 11—and the first rule of sanitation—concerns resistant varieties. Plant breeders like to aim at high resistance or complete immunity; this is desirable, and the effect of immunity is readily understandable: there is no epidemic (and in that case the matter would be outside the scope of this chapter). But commonly the available resistance is only partial; infection occurs, but at a slower rate. The magnitude of r is reduced, and the value of sanitation correspondingly increased. Every quantum of resistance, however small, increases the efficiency of sanitation; and every quantum of sanitation increases the value of the partial resistance. As methods of control, sanitation and partial resistance should go together. When a disease is controlled by sanitation, partial resistance in any measure is an achievement not to be despised. It has been much undervalued in plant pathology, largely as the result of ignoring its connection with sanitation.

E. Secondary Epidemics

An epidemic can be defined as secondary if it starts from inoculum derived from another, earlier epidemic. In Section I B it was observed that epidemics of eastern X disease of peaches, Pierce's disease of grapevines, and tomato spotted wilt on tobacco are secondary because there is apparently no spread of the diseases in the crops mentioned. Infection must come from outside. In the Netherland van der Zaag (1956) found that with the potato variety Duke of York there was about 1 primary focus of infection per square kilometer at the start of the season, this focus originating from an infected seed tuber. With the more resistant variety Noordeling, primary foci were much rarer or entirely absent. Further, he tested varietal differences by artificially inoculating leaflets in the field before natural infection was apparent, and then 12 days later counting how many leaflets had become secondarily and naturally infected from these primary artificial sources.

For every 1 lesion on Noordeling that developed secondarily around the primary source, there were 180 on Duke of York. From this we infer that the magnitude of r was roughly 180 times as great for Duke of York as for Noordeling. This estimate is rough but adequate; the evidence does not allow an exact estimate of either r or r'. Epidemics in Duke of York take about a month to develop from the primary foci, so, from equation 11, an epidemic in Noordeling would need 180 months to develop. If one adds to this the information about the scarcity of natural primary foci with Noordeling, the period needed would be much greater. Even if one admits the possibility of error, the differences are striking enough to make it reasonably certain that a primary epidemic of blight in Noordeling is improbable in a growing season limited to 4 or 5 months. If Noordeling becomes significantly blighted, it is almost certainly as a result of infection obtained initially from some other variety. The tendency for disease to spread from more to less susceptible varieties and cause secondary epidemics in them, increases with the extent to which initial inoculum must multiply to cause an epidemic. For any given time up to the onset, the logarithm of the amount of multiplication is proportional to *r*. If two varieties differ in that in the one *r* is twice as great as in the other at all times, then by the time the more susceptible variety has multiplied 10^2 times the less susceptible will have multiplied 10 times; by the time the more susceptible has multiplied 10^6 times the less susceptible will have multiplied 10^3 times. As multiplication continues, the difference in level of disease between the varieties increases.

The more the multiplication needed before a level of, say, 5% is reached, the greater is the difference in disease between the varieties when the more susceptible has reached that level; hence, other things being equal, the greater the chance that disease in the less susceptible variety will be influenced by inoculum from the more susceptible. With diseases like potato blight, in which relatively little inoculum survives the winter, the danger of susceptible varieties starting secondary epidemics in other varieties is correspondingly great. Late blight, caused by *Phytophthora infestans*, can survive perennially on tomatoes in countries in which tomatoes are grown all the year round. It can also be brought into the summer-grown tomato plants of colder eliminates by blight-infected transplants imported from warmer areas.

But late blight of tomatoes has also a considerable literature of epidemics secondary to those of potatoes. Neighboring blighted potato fields or potato refuse piles are repeatedly implicated as the source of tomato blight. But in one detail there has been a noteworthy change in the literature since Mills (1940) summarized it. In 1940 races of *Phytophthora infestans* from potato were poorly adapted to tomatoes, and according to Mills, had to be trained to attack tomatoes by a few passages through tomato leaves. Nowadays there are frequent references in the literature to potato races that readily attack tomatoes, although a distinction between potato and tomato races still exists. On the general evidence of the literature, secondary epidemics on tomatoes are now more easily established from potatoes. There has been, it would appear, a change in the relationship between *P. infestans* and the potato, which has also involved the tomato secondarily and possibly incidentally, and which has perhaps been the cause of the destructive epidemics of tomato blight on a scale unknown before 1946.

There is a type of secondary epidemic brought about the mixed cropping. The apple variety Delicious has considerable resistance to powdery mildew, caused by *Podosphaera leucotricha*. Jonathon is very susceptible and is commonly used as a pollinator for Delicious. In mixed orchards spores from diseased Jonathans may start secondary epidemics on Delicious which are severe enough to require control measures. Harwoods grown alone are normally resistant to *Fomes annosus* but may be killed when grown in mixed stands with susceptible conifers in which inoculum builds up. These examples show the value of uniformity of resistance in minimizing the risk of secondary epidemics, a point to remember when reading Sections V, C, 3, f and V, E.

Secondary epidemics provoked by mixed cropping are often used by plant breeders to eliminate susceptible lines of seedlings. A variety known to be susceptible is interplanted with the material to be sorted out in order to ensure that infection will be present in the breeders' plots. The method is simple and convenient, but there is the danger of confusing the smaller amount of resistance adequate to stop a primary epidemic with the larger amount needed to cope with a secondary epidemic.

III. The Spread of Epidemics

A. Factors That Affect the Rate of Spread

Section III describes how an epidemic spreads over a distance. Three factors are considered at the start. A fourth factor, scale of distance, is discussed later in the section.

1. *The Gradient of Infection*. The probability that a healthy plant in a given direction will become infected depends on the distance of that plant from the source of infection. Infection grades away (usually smoothly, but not necessarily so) from the source. Unless otherwise stated, the source will always be taken to be a point--a singly plant, for instance—and not a field or strip. Gradients should be determined only when the percentage of infection is low; apparent gradients necessarily become flatter as disease mounts.

2. *The Abundance and Distribution of Susceptible Host Plants*. The probability that a spore or other propagule will travel a given distance is proportional to the number of propagules released at the source. If the number of air-borne spores at the source is multiplied 100 times, then on an average 100 times as many as before will blow

past the first milestone, 100 times past the second, and so on. When heavy spore showers are detected hundreds of miles from the nearest source, it is reasonably certain that the source is not a single plant but fields of plants, not just a few fields but hundreds of acres of fields. Abundant host plants mean, too, that the leaps the pathogen need make from one host plant to another are small. Their distribution—whether in small fields or large—also affects the pathogen's spread; this is a matter to be discussed in Section IV.

3. *The Rate of Multiplication*. Spread is multiplication at a distance from the source of inoculum; and, for a given gradient, multiplication at a distance is related to multiplication in general. Fast multiplication of disease lesions means fast spread of disease. The connection has long been tacitly accepted, and pathologists commonly use the words "multiplication" and "spread" interchangeably. A tie between multiplication and spread is discussed in Section III, E (equation 13).

B. Horizons of Infection

Horizons of infection have been discussed by van der Plank (1949). For argument's sake suppose temporarily that the probability that a lesion (or, for systemic diseases, an infected plant) will form a daughter lesion in unit time on a healthy plant at a distance s is

$$p = \frac{a}{3^b} \tag{12}$$

Here a and b are constants. We shall not bother about the accuracy of this equation; the relationship shown is just scaffolding, which will be removed later. Suppose that there are infected fields scattered over a large area, and consider any field as center. Suppose that this field is only lightly infected: that the epidemic in it has not gone further than the onset, and that it recieves inoculum from other fields in a sector narrow enough for the gradients toward then center to be considered uniform. The probability that a lesion two units of distance away in the sector is $1/2^b$ times the probability that a lesion will develop from inoculum from a parent lesion one unit of distance away in the sector. But on an average the number of parent lesions two units of distance away is twice the number one unit away (because for a given angle at the center the arc is proportional to its radius). Hence on an average the number of lesions caused by inoculum received from all sources (all parent lesions) two units away is $2/2^b$ or $1/2^b$ times the number from inoculum received from all sources one unit away. Similarly the number from inoculum received from all sources three units away is $1/3^{b-1}$ times the number from all sources one unit away. If Q is the number of lesions from inoculum received from all sources one unit of distance away the number of lesions from inoculum received from all sources at all distances is

$$Q\left(1+\frac{1}{2^{b-1}}+\frac{1}{3^{b-1}}+\ldots\right)$$

If $b > 2$, this series is convergent, and there will be a limit from beyond which inoculum will not come.

To simplify calculations, suppose that the distance between fields is relatively large and that the fields are uniformly infected and uniformly distributed. We can, as an adequate approximation, take the distance between the centers of neighboring fields as the unit of distance. If $b = 2.5$, 62% of the daughter lesions caused by inoculum received from other fields will come from fields, beyond the immediate neighbors, 40% from more than 3 fields away, and 32% from more than 5 fields away. If $b = 3$, the respective figures are 39, 17, and 11%; if $b = 4$, the figures are now 17, 3, and 1% respectively. If, for example, one regards inoculum coming from behind the horizon as negligible if it is responsible for less than 10% of the daughter lesions, then if $b = 2.5$, a horizon is established more than 50 fields away; if $b = 3$, about 5 fields away; if $b = 4$, 2 fields away. The horizon draws in sharply as b increases, i.e., as gradients become steeper.

It is not assumed that gradients are the same in all sectors, and the horizon about a field need not be circular.

C. Continuous and Discontinuous Spread of Epidemics

If $b > 2$, the source of a new outbreak will probably be within a horizon. The greater the magnitude of b, the more likely is the source to be near, and the easier it will be to follow the path of an epidemic. The spread will be continuous. But if $b < 2$, more inoculum will arrive from far than from near (assuming of course that host plants occur over a wide area). Infection will appear as if "from nowhere." There will be what have been called "spot" infections—infections that cannot be traced to their source. The spread will be discontinuous. Low value of b can be expected if the movement of inoculum is oriented with a restriction on random scattering, as would occur, for example, if the inoculum were carried by birds migrating toward a particular destination. Migratory birds are thought to spread chestnut blight by carrying the sticky pycnospores of *Endothia parasitica*. "Spot" infections were a feature of the great chestnut blight epidemic in North America, and occurred in addition to local infections caused by wind-blown ascospores; it seems likely that during migrations b was less than 2.

D. The Determination of Gradients; the Scrapping of Equation 12; Dutch Elm Disease; Potato Blight

Equation 12 was used only to build up an argument. In scrapping it one need not replace it by a better equation, but only consider its most useful features. The central inferences from the equation concern the value $b = 2$, or, to scrap the equation, concern a gradient in which the number of lesions formed by inoculum from a source varies inversely as the square of the distance from the source. If one plots the number of lesions against distance on a log-log scale, the inverse-square lines are straight. One can draw as many as one wishes—all parallel. Such are the lines *A, B, C, D* in Fig. 6.3. The observed

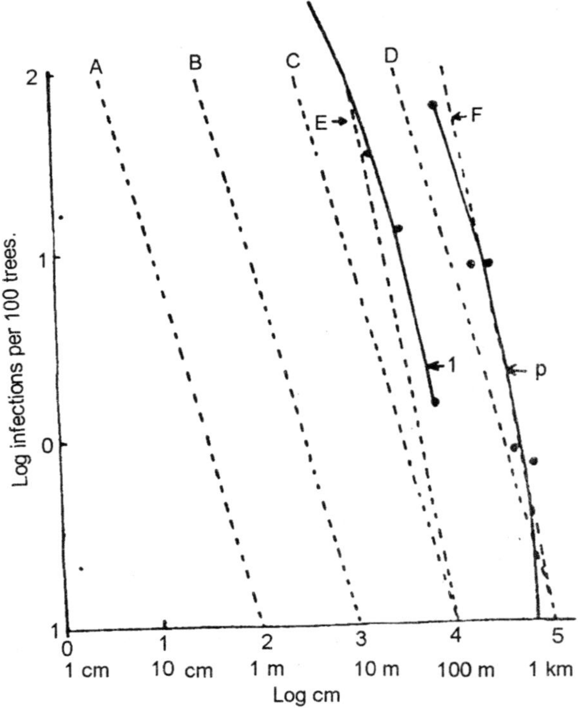

Fig. 6.3. Amount of Dutch elm disease (Ceratostomella ulmi) at varying distances from the source of inoculum. Curves 1 and 2 respectively, shown as transformed numbers of infections. A, B, C and D are inverse square lines with scales of distances as 1:10:100:1000. E and F are inverse-cube lines with scales of distance in the ratio 1:10.

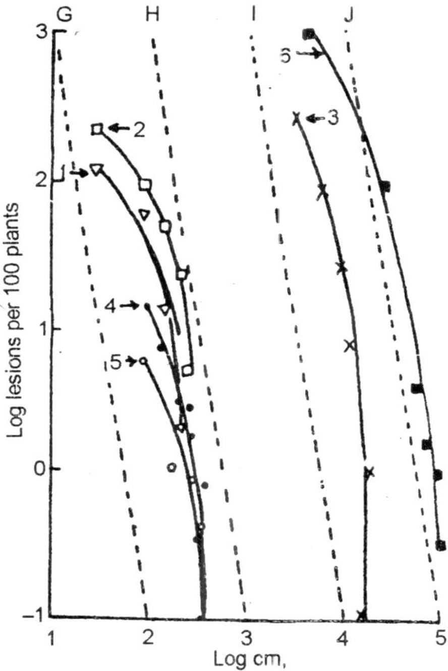

Fig. 6.4. Amount of potato blight (P. infestants) at varying distance from the source of inoculum.

gradient for any disease can then be compared immediately with these lines. If its slope is steeper than theirs, there will be a horizon of infection, and spread will be continuous. Similarly one can draw a number of parallel inverse-cube lines like the lines *E* and *F* in Fig. 6.3. If the observed gradient is steeper than theirs, horizons can be expected to be fairly close. In Fig. 6.4 the process is continued, and lines *G, H, I,* and *J* show the number of lesions inversely as the fourth power of the distance from the source of infection.

There is a fair amount of information about it. Among other things, the spread of disease from a single infected tree in a limited period of time has been observed by several workers to virtually cease within some hundreds of yards from the tree, so one knows in advance that one should expect gradients steeper than the inverse-square lines. On one point there is a difficulty—a difficulty common to most of the literature of diseases that could be used for illustration: gradients should be determined only at low percentages of disease. This limits information to the lower part of the curve; the disadvantage of this is that information there is usually based on relatively few diseased plants and is consequently not powerful statistically.

At higher levels of disease one can correct the curve partially (or fully, if the lesions are distributed randomly) by transforming percentages of diseases into calculated numbers of infections per 100 plants. Attention to this was drawn by Gregory (1948), who published a useful table of transformed values. Some examples will explain the transformation. Suppose there were exactly 100 random infections per 100 plants. Not all plants would be infected; on an average, 36.8% would have no infections (they would remain healthy) and 63.2% would have 1 or more infections. The table calculates this in reverse; it transforms 63.2% disease to 100 infections per 100 plants. The highest figure for disease is 88.9%, which is transformed to 220 infections per 100 trees. At low percentages the change is small and usually negligible: 5% disease transforms to 5.13 infections per 100 plants.

The combined data of Zentmyer et. al. (1944) for three plots in Connecticut; and the data of Liming et. al. (1951) for a plot in New Jersey. These data were collected within a year of the emergence in large numbers of the vector beetles from the source of infection which in each case was a single naturally infected tree. There was therefore little time for secondary multiplication to affect the gradients. At the lower levels f disease, which are the safest to observe, the gradients are steeper than the inverse-square lines and are more like the inverse-cube lines. This is what one would expect from general observations on the disease, but the data are too few to warrant detailed statistical analysis.

Analyzes records of potato blight, caused by *Phytophthora infestans*. The data of Limasset (1939) were taken from the paper of Gregory (1945). Bonde and Schultz (1943) recorded the number of lesions per 100 plants, so no transformation is needed. These records were taken from a field about 100 ft. away from an infected dump pile on June 12, and had infected most of the plants in the pile by June 25. The readings in the potato field that are reproduced were taken on July 12. It is thus possible that there had been some secondary multiplication in the field before then, so the gradients, steep as they are, may underestimate the true steepness. The results of Waggoner (1952) in curve 4 were obtained 9 days after artificial inoculation of the center from which sporangia later spread; those in curve 5 from another plot18 days after inoculation, during 9 days of which the weather was generally unfavourable for infection.

In addition, data are presented for the related fungus *Peronospora destructor* to make the records as complete as possible. They are derived from the data of Newhall (1938) for downy mildew lesions per 100 ft of row, and no transformation was needed. These various records are consistent in suggesting that, as a reasonable approximation at the low amounts of disease in which we are primarily interested, the number of lesions falls off inversely at least as the fouth power of the distance from the source. The result has been assumed in the calculation made for blight in Section II, C. Two more curves are shown for *Phytophthora infestans*. They are slightly steeper than an inverse—sixth-power gradient. The evidence from all the records that the curves at low amounts of disease become steeper than an inverse—fouth-power gradient is statistically significant; but it is inadequate to say how steep they eventually become. Curve 3 in Fig. 6.4 is displaced to the right of curves 4 to 5. Displacements will be discussed under the next heading. But one reason for displacement seems evident here: the source of infection for curve 3—the cull pile—was comparatively large and heavily infected, and was at a distance from the field; the sources for curves 4 and 5 were smaller and within the fields.

E. Scales of Distance; a Tie between Multiplication and Spread

For illustration let us return to the chestnut blight fungus spread by migrating birds. There are two relevant features about birds during migration: the flights are longer between pauses, and are oriented toward some particular destination. The second feature—orientation—was cited as probably involved in making spread discontinuous. The first feature—long flights—was ignored. Making flights longer without change of orientation and without serious loss of inoculum during flight would make an epidemic spread on a larger scale (in the word's literal meaning of relative dimensions). But it would not make a continuous spread discontinuous. Consider the inverse-square lines in Fig. 6.3.

The scale of distance for line *D* is exactly 1,000 times the scale of line *A* at any given level of infection. Between any two given levels of infection the change from line *A* to line *D* is a change from millimeters to meters, from meters to kilometers. In Fig. 6.3 displacement to the right means an increase of scale of distance. Other things (including the strength with which the inoculum is emitted at the source) being equal, longer flights by winged vectors or longer motions by the inoculum generally mean a larger scale of distance and displacement to the right. Restricted motion, which one might expect of inoculum moving through heavy soil, would mean a smaller

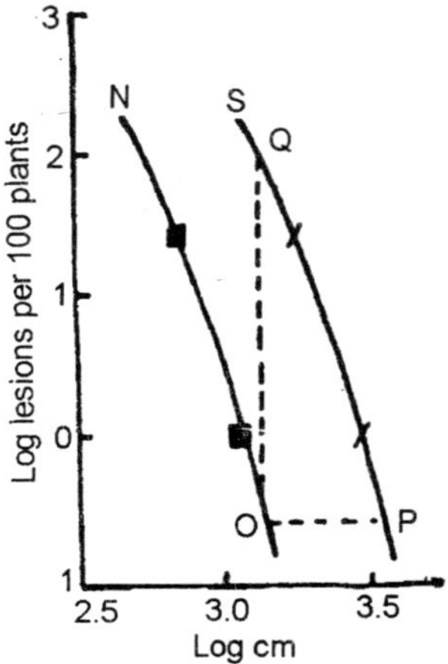

Fig. 6.5. Amount of potato blight, P. infestans, at different distances from the source of inoculum.

scale and displacement to the left. When disease spreads from a single point source of inoculum, one can define the scale of distance on a relative basis by saying that the scale of distance varies directly with the distance between two given levels of disease (which must be defined if necessary). When curves relating disease and distance are plotted in a log-log graph and the displacement of the curves is fairly regular or when the two levels of disease are selected close together, a change of scale of distance is readily determined graphically. Consider Fig. 6.5, which is constructed from data of van der Zaag (1956) for the spread of *Phytophthora infestans* from an incompletely removed source of infection.

The data for percentages of diseased plants have been transformed into the number of lesions per 100 plants. To the north of the focus (curve *N*) the spread was less than to the south (Curve *S*). Since the spread in both directions was from the same source of inoculum, the displacement of curve *S* to the right of curve *N* is the result of a changed scale of distance. This is measured by the line *OP*, which represents an increment of 0.4 in the logarithm of the distance or a 2.5-fold increase in scale of distance at the level of disease at or near points *O* or *P* (log 2.5 = 0.4). The argument will have been grasped in the simpler comparison between the straight lines *A* and *D* a few paragraphs back, and needs no further elucidation.

The corresponding change in the scale of disease at the distance represented by points *O* and *Q* is measured by the line *OQ*, which represents an increment of 2.5 in the logarithm of the number of lesions or a 316-fold increase in the number (log 316 = 2.5). At this distance there was about 316 times as much disease to the south as to the north as a result of a 2.5-fold change of scale of distance.

The records of Waggoner (1952) for *P. infestans* in his plot at Clear Lake in 1950 on day eighteen show a scale of distance to the NNW about 8 times as great as to the SSE. Since, approximately, disease varied inversely as the fourth power of distance from the source, there was about 8^4 or 4096 times as much disease over a given distance to the NNW as to the SSE. This result appears to have been due to wind. There is considerable evidence on the literature that wind can strongly alter the scale of distance. One reason why wind affects the scale in different directions is that more inoculum leaves the source in one direction than in another.

To determine a relation between multiplication and spread, referred to in Section III, A, 3, consider a primary gradient set up in healthy plants surrounding a point source of infection. A primary gradient is one set up by inoculum derived directly from the initial point source without complications from secondary multiplication of disease in the zone of the gradient. For example, curves 1 and 2 probably represent primary gradients of Dutch elm disease. Return to equation 7 and consider a boundary condition. While $t < P$, the quantity x'_{t-P}, taken here as the point source, is constant. Because the gradient is set up in plants healthy at the start, i.e., when $t = 0$, the amount of disease at any given time less than P and (for a given gradient) at any given distance from the source, will be proportional to the value of r' prevailing over the period of the observations. Whence, to follow the reasoning in the previous paragraphs,

$$\Delta s = kr'^{(1/b)} \tag{13}$$

where Δs, which determines the scale of distance, is the distance between two given levels of disease in the gradient, k is a constant; b has the same meaning as in equation 12; and changes in r' are assumed to be moderately small.

If the two levels of disease are taken fairly close together, the value of b can, with reasonable accuracy, be taken as constant over Δs. This feature appears in all the curves, especially at distances not too near the source. Therefore the use of b here does not revive equation 12, but only the part of it that may legitimately by revived. It is necessary that b should be positive; i.e., there must be an actual decrease of disease with distance away from the source.

Equation 13 holds for those factors (except the incubation period) which effect r but not the gradient. With such a range of factors affecting r—from wind to the susceptibility of the host—and such a range of inoculum—from air-borne spores to water-splashed bacteria and vector-transmitted viruses—any generalization about the effect on gradients is difficult. It is, however, worth noting that in their effect on scale of distance the factors that determines r cannot be distinguished one from another (expect the incubation period). Consider, for example, diseases that spread mainly "by contact" between plants that are immediate neighbors. It is common experience that these diseases multiply slowly, and it is generally assumed that they multiply slowly because inoculum can only be spread by contact. In many cases one can use the opposite argument: the diseases appear to spread by contact because they multiply slowly for reasons that need not be known.

A summary of the basic concepts might facilitate an understanding of the relation between multiplication and spread. We have been concerned with a point source of inoculum, hence only with primary gradients, hence not with the incubation period, and hence with r' and r, as equation 13 shows. If inoculum comes from a source other than a point (e.g., an epidemic's front) the incubation period and r may become involved. A particular solution for the relation between r and the spread of the front of an epidemic of wheat stem rust underlies the Appendix. It is for the special case in which the pathogen's birth rate greatly exceeds it death rate and lesions are small.

F. Inverse Scales of Distance; the Law of Lesion Size

We have been considering changes in the distance scale on which a pathogen can spread. Now let us think of the problem form the opposite view ad consider changes in the scale of distance over which a pathogen must spread if it is to transmit disease. As an example we may consider the law of lesion size. The law is that the highest potential rates of multiplication are with small lesions. A systemically infected plant is regarded as a single lesion extending over the whole plant. (The largest lesion is a systemically infected tree). In round figures, small local lesions like those of potato blight can multiply one-billionfold in a season, systemic diseases of small herbaceous plants up to ten-thousandfold, and systemic diseases of trees about tenfold in a year. Evidence is given elsewhere. The word "law" is used to indicate that the relationship was observed, not predicted. Only maximum rates are relevant; it is not relevant to know that in arid climates or immune varieties potato blight does not multiply at all.

Two virus diseases with efficient aphid vectors are cauliflower mosaic and tristeza disease. The number of cauliflower plants infected with cauliflower mosaic multiplies faster than that of orange trees infected with tristeza disease. Suppose cauliflower plants in a field could be enlarged to 10 times their normal size with all distances increased in proportion: the plants would be 10 times as tall, 10 times as broad, with the distance between them 10 times as great, and in rows 10 times as wode. Suppose one plant in the field is infected. The distance from any relative position on this plant to any relative position on a neighboring plant or any relative position on any plant in the field would be increased 10 times, by the process of enlargement. The scale of distance that an aphid would have to travel to transmit the virus would be increased 10 times, and the probability of a transmission would drop, correspondingly, according to the gradient.

Similarly, if orange trees in an orchard could be shrunk to one-tenth their size with all other distances shrunk proportionately, the probability of transmission would rise in accordance with a decrease to one-tenth of the scale of distance over which an aphid must travel. So, other things being equal, an aphid as a greater chance of transmitting mosaic between cauliflower plants than between orange trees simply because orange trees are larger. The argument applies not only to vector-transmitted viruses but to any sort of inoculum. It applies not only to systemic disease but to any lesion; a uredospore of a cereal rust fungus must travel over only a very distance, a few millimeters, to fall clear of an isolated pustule on healthy tissue. Cereal rust pustules can multiply fast because the scale of distance the pathogen must travel is small. Inverse scales of distance are only one of the factors involved in the effect of size of lesion. The incubation period (to mention another) is often longer in systemic disease; the effect of this period has already been given in equation 8. Forest trees seem to have resistance to virus disease, systemic bacterial diseases and systemic smut diseases. Operating in the opposite direction is the factor of the size of plant (Section VI, D, I).

IV. Epidemics in Relation of the Abundance and Distribution of Host Plants

A. General

Agriculture has meant bringing plants together in fields and—to meet the ever growing demand for food—increasing the acreage of these fields. This raises two separate questions with which this section is concerned. First: how does the bringing together of plants and increasing the acreage affect the prevalence of disease? The general answer is that disease is encouraged. But there are exceptions, and increasing the acreage, if it means enlarging the fields without increasing their number, may at times even reduce the percentage of disease. Also, apart from this exception, the relation between disease and intensified agriculture is not simple. Some diseases are more sensitive to intensified agriculture than others, and increase relatively faster as the acreage is increased.

The second question is: how does the distribution of host plants affect epidemics? For example, if the acreage remains constant, how do the size and distribution of the fields influence infection? The answers can be of the greatest practical importance if epidemics can be reduced by nothing else than planned agriculture. The least exploited method of reducing plant disease is by planning the pattern of farming in directions other than by crop rotation; no attempt is made here to produce a blue-print for such exploitation, but it is hoped at least to indicate that blue-prints are entirely feasible.

B. Disease in Susceptible Plant Scattered among Immune Plants

Although bringing plants together into fields normally increases the likelihood of disease, epidemics among scattered host plants are not uncommon. For example, *Puccinia malvacearum* attacks only wild and garden plants. Yet it spread through western Europe within 5 years of its first appearance. But bringing plants together may sometimes necessary condition for an epidemic to start. A good example is provided by the "native" potatoes of the mountains of Basutoland. The chief vector of the potato viruses, *Myzus persicae*, is abundant, and its winter host, the peach, almost ubiquitous.

The varieties have no great resistance to virus disease, and succumb quickly when they are planted in garden plots. But when they grow—as they ordinarily do—as odd plants in the corn fields or in the grass near roadsides, there is little evidence of virus disease, and the varieties have persisted for more than a century in good health. Mixed cropping has been deliberately used to control disease. Planting squashes and marrows in corn is commonly practiced in South Africa to control an aphid-borne mosaic disease. In Denmark and England beet seedlings intended for seed crops have successfully been protected against virus yellows disease by shield crops of barley. In France the idea has been extended. Seedlings of beets are raised in beds which are latticed with longitudinal and cross bands of barrier crops to give subdivisions about 20 yd. long by 5 yd. wide. The barrier bands are about 1 yd.

wide. Oats, corn, hemp, and sunflower are generally used. In England, Broadbent (1957) showed that narrow barriers of barley rows reduced the incidence of cauliflower mosaic and cabbage black ring spot in cauliflower seedbeds, a matter that has already been mentioned in Section II, A as an illustration of a quantitative problem in sanitation. There is a considerable range in detail as one passes from mixed cropping through shield crops to barrier crops; but all are methods based on the reduction of the effective movement of inoculum, and all share the common feature that they are useful in practice only against a type of disease which we shall call a crowd disease, i.e., a disease that is likely to reach epidemic proportions only when the host plants are crowded together into fields.

C. The "Epidemic Point" in Crowding Plants Together

Before crowd diseases are discussed consider what happens when susceptible host plants are crowded together. It is useful here to return to our introductory concept of epidemics as a matter of balance between the birth and death rates of the pathogen. Bringing plants nearer together increases the movement of inoculum between them; and it increases the birth rate of the pathogen. Bit it is unlikely to affect the death rate greatly (unless the pathogen or its vector is itself parasitized or preyed upon). When the point is reached at which births exceed deaths the danger of an epidemic beings.

Take as examples three disease of trees: rosette disease of peaches, swollen shoot of cacao, and tristeza disease of sweet oranges on sour orange rootstocks. All are caused by viruses that kill their hosts and die with them. Rosette virus ordinarily spreads slowly from peach to peach, and its birth rate is low. However, its death rate is high, because it quickly kills its peach host, which usually dies in the same season in which symptoms are first seen, and then dies too. Ordinarily, the death rate exceeds the birth rate and the disease in peaches is self-eradicating although it occasionally flares up and affects a whole orchard. Swollen shoot of cacao has the same features as peach rosette, but not in quite such an extreme degree. It spreads slowly, but not quite so slowly; it kills quickly, but not quite so quickly, and infected trees may survive up to 2 years or longer.

Tristeza disease has an abundant and efficient vector, *Toxoptera citricidus* (among others), which occurs in both winged and wingless forms, and infection spreads relatively fast through an orchard in which *T. citricidus* occurs. The disease kills sweet oranges on intolerant rootstacks but not very fast; the trees go into a decline that may last for several years. Compared, then, with peach rosette, citrus tristeza has a higher birth rate and a lower death rate. The tree diseases from an interesting series with regard to the epidemic point, which we define as the point in crowding plants together at which the birth rate of the pathogen exceeds the death rate. In peach rosette the epidemic point is usually imaginary: it is usually not reached even when peaches are in continuous orchard formation. The disease of peaches is usually sporadic and self-eliminating, and the virus persists only because it has other more tolerant hosts such as the wild plum, from which it spreads to peaches. If epidemiology is the science of disease in populations, inpeaches this behavior is epidemiological hypersensitivity, comparable in its effect on populations with ordinary hypersensitivity in its effect within the plant. Swollen shoot of cacao has a lower epidemic point, which was apparently reached in West Africa between the World Wars, when cacao production was expanding rapidly. Before then the disease in cacao had a long history of sporadic outbreaks; it was only afterwards that the epidemic began to rage.

Tristeza seems to have a fairly low epidemic point in countries in which *Toxoptera citricidus* is present; in these countries the virus seems to pervade quickly all susceptible species of citrus, killing those trees which are on unsuitable rootstocks and saturating the rest. It is reasonable to suppose that with all diseases that cause primary epidemics the host plants are crowded beyond the epidemic point, at least duri ng epidemic years. (The epidemic point will necessarily be lower in years favorable to the disease). If the epidemic point is passed during the recorded history of man's crowding plants together, there will be a discontinuous change from sporadic disease to epidemic disease, as has happened apparently with swollen shoot of cacao. But where the epidemic point is low it was probably passed long ago, and the process of intensifying agriculture will seem to have a continuous effect—with epidemics becoming continuously more likely to occur.

D. Crowd Diseases; Changes in the Relative Importance of Disease; a Glimpse of the Future

Crowd diseases are likely to reach epidemic proportions only when the host plants are crowded together; they are diseases with a high epidemic point. The inoculum is likely to have a high death rate (as when, for example, it cannot persist in the soil); or the movement of inoculum between separated plants is likely to be small because of a small scale of distance, as reflected in a steep gradient and a low value of r or, more strictly, r'. By equation 11 the effect of variations in the amount of inoculum reaching a field is greatest when r is least. To make a long story short, consider only diseases like potato blight, the cereal rusts, and those virus diseases not transmitted directly through the soil or dead plant residues.

Usually, the systemic virus diseases multiply more lowly than the local lesion diseases. Therefore, intensified agriculture, by bringing fields nearer together and increasing the movement of inoculum between them, is likely to advance epidemics of systemic virus diseases relatively faster than those of local lesion diseases. To some extent the trend has been observable during the past century of fast rising food production. Plant pathology started as a branch of mycology, but became increasingly more concerned with virus diseases. One grants that much of this history has been fortuitous; had de Bary's chief interest been in virus diseases, the history would have been different. But the change was not wholly fortuitous. Further, because diseases especially sensitive to the intensification of agriculture are, as a result of this characteristic, especially amenable to control by isolation and sanitation generally, plant pathologists can expect to be much more concerned with the theory and practice of sanitation in the future than in the past.

E. The Effect of Making Fields Larger

1. *Inoculum Coming from Other Hosts*. Consider the finding of Wellman (1937) in Florida that cucumber mosaic virus, causing mosaic in celery, comes from other hosts such as *Commelina nudiflora* growing within 75 ft. or of vectors flying right over the field without settling in it. The area of 75 ft. of the field. For the sake of argument, accept this figure and ignore the possibility of some inoculum arriving from beyond 75 ft zone about a square field of 0.1 acre is 0.86 acre; about a field of 1 acre it is 1.85 acres: about a field of 10 acres it, it 4.95 acres. Thus 10 acres of celery divided into 100 square fields of 0.1 acre, would draw infection from other hosts scattered over 86 acres; divided into 10 square fields of 1 acre, from other hosts scattered over 18.5 acres; and concentrated into 1 square field of 10 acres, from other hosts scattered over only 4.95 acres. Concentration of the host plants into a single field of compact shape reduces infection from outside to a minimum. Probably one of the few occasions on which the percentage of disease is reduced while agriculture is intensified is when infection comes from scattered hosts near the fields and acreage is increased by increasing the size of the fields and not their number.

2. *Inoculum Moving between Fields*. We shall consider here what happen when fields are made larger and correspondingly fewer, so

that the total acreage remains constant. To do so we shall revive equation 12, but only to the limited extent that we use it, as in Section III, D, to illustrate the gradient of disease. Discussion will be limited to what happens before the onset of a epidemic,i.e., to inoculum moving into fields that are not already heavily infected, which is the only case of practical importance. Assume that the fields are equal in size and uniformly shaped, orientated, and distributed over the country. Making them larger and correspondingly fewer increases the average distance between them in proportion to the square root of their average area, their shape and orientation being unchanged.

The average distance, not necessarily on a straight course, that a spore (or other propagule) must travel from any particualr relative position in the field of its origin to any particular relative position in the first field it reaches is proportional to the square root of the average area of the fields, that is, it must travel an average distance $k_1A^{1/2}$ where k_1 is a proportionality constant and A the average area of the fields. The probability that a lesion (or, with systemic diseases, an infected plant) in one field will form a daughter lesion (or infected plant) at any particular point in the first field, is, therefore, by equation 12

$$\frac{a}{k_1^b A^{\frac{1}{2}b}}$$

The average distance the spore must travel to cross the first field, that is, the average potential number of daughter lesions along its track across the field, is also proportional to the square root of the average area. The probability of a daughter lesion somewhere along the track can therefore be taken as

$$\frac{ak}{k_1^b A^{\frac{1}{2}(b-1)}}$$

where k is another proportionality constant. Similarly, if the spore passes the first field, the probability of a daughter lesion in the second field it reaches is

$$\frac{ak}{k_2^b A^{\frac{1}{2}(b-1)}}$$

in the third

$$\frac{ak}{k_3^b A^{\frac{1}{2}(b-1)}}$$

and so on where k_2, k_3...... are constants. The probability of a daughter lesion in any field other than the field of origin is:

$$p = \frac{ak}{A^{\frac{1}{2}(b-1)}}\left(\frac{1}{k_1^b} + \frac{1}{k_2^b} + \frac{1}{k_3^b} + \ldots\right)$$

The series

$$\frac{1}{k_1^b} + \frac{1}{k_2^b} + \frac{1}{k_3^b} + \ldots$$

may be taken as convergent when $b > 1$ because it is almost identical with the series

$$\frac{1}{k_1^b}\left(1 + \frac{1}{2^b} + \frac{1}{3^b} + \ldots\right)$$

which is convergent when $b > 1$. Hence, when constants are collected together as K,

$$p = \frac{aK}{A^{\frac{1}{2}(b-1)}} \qquad (b > 1) \tag{14}$$

within the limit of the initial assumption of uniformity.

As an example of the effect of area, with an inverse-fourth power gradient (which on present data seems a fair approximation to the behavior of *Phytophthora infestans* not too near its source), doubling the average area of fields and halving their number reduces the probability that a lesion will cause the development of a daughter lesion in some other field by 65%; trebling their area reduces the probability by 81%. If one wishes to show that the number of daughter lesions in other fields follows the same trends, it must also be shown that the multiplication of disease is the field of origin is not appreciably affected by the area of this field. This follows from direct observation. Increasing the area of fields reduces the amount of inoculum that escapes from the field of origin. If this affected the multiplication of disease there, one would expect gradients of disease within the field, the border being significantly less diseased than the center.

Small border effects have indeed been observed: e.g., Thomas et. al. (1944) found that peach yellow-bud mosaic occurred less in the outside row or two, when disease multiplied within and the orchard was not exposed to infection from without, and Storey and Godwin (1953) found that when cauliflower mosaic multiplied within a field the incidence was somewhat less in the outer rows. But these border effects are small and extend inward for only a few rows or feet. If they had been strong, they would long ago have received general comment, and one may infer that with the exception of very small plots and fields, the area of the field does not greatly affect the rate of multiplication within. The reason is not difficult to find. The vast bulk of inoculum released from a field falls back into the same field. Variations in the small proportion that escapes have a large effect on inter-field movement of disease, but very little on disease within the field of origin itself.

F. The Paradox; in Praise of Large Fields; the Second Rate of Sanitation

The paradox is this. Bringing plants together into fields increases the chance of epidemics; bringing them still further together, by increasing the area of the fields and correspondingly reducing their number, may reduce the chance of a general epidemics. In the literature of plant pathology it is common to read of the danger of bringing together the host plants of a pathogen. The indiscriminate indictment for bringing plants together is unjustified. Food must be grown, and it is time to write in praise of the large field. Take these two examples: swollen shoot of cacao can be controlled by sanitation, i.e., by cutting out diseased trees. But the scope for this on small farms is limited, and the difficulties have been described by Posnette (1953). "Each farm, consisting usually of less than 5 acres, has to be treated separately although it has no clearly defined boundaries, and from the standpoint of disease control about 50 acres is the smallest area it is practicable to consider as a unit. For treatment to have any permanent value, the unanimous co-operation (or at least the consent) of many individual farmers must be obtained, and no method of achieving this has been found. Consequently, the Ivory Coast, Nigeria, and the Gold Coast [Ghana] have each in turn been forced to abandon their original plants for disease eradication and have had to adopt the expedient of a 'cordon sanitaire' around the heavily infected districts."

At the other end of the scale we read of the trend gardening has taken in the Everglade region of Florida in recent years. Anyone farming less than a section or two of land—1 or 2 square miles—is considered a little man in the vegetable patch. Celery fields are half a mile long. A celery harvester has been described that weighs 60,000 pounds and carries a crew of nearly 100 persons abroad. Weeds are cleaned our chemically. The writer does not pause to mention such a trifle as disease, but if our inference from Wellman's findings is correct (in Section IV, E, 1) celery mosaic from weeds must cease to exist as a practical problem.

If inoculum comes from other hosts—as cucumber mosaic virus in celery fields comes from *Commelina* and other weeds—making fields larger reduces the percentage of infection in the fields. If inoculum moves from field to field, making fields larger and correspondingly fewer reduces the movement. If sanitation is practiced within the field—as by using clean seed if the pathogen is seed-borne, by rotation of crops if the pathogen persists in the soil, by roguing out diseased plants, and by planting disease-free nursery stock—the size of field does not affect the process directly. But making fields larger and correspondingly fewer affects it indirectly by reducing the reentry of inoculum from without. Guarding against the reentry of large amounts of inoculum is part of the process.

One can summarize all this in a second rule of sanitation: efforts at sanitation are furthered by making homogeneous fields, orchards, and plantations larger and correspondingly fewer. The qualification that the fields should be homogeneous is implicit in all our arguments. If, for example, part of a large field is sown with healthy and part with diseased seed, there is no necessary advantage in largeness; it would have been better to separate two parts. Sanitation needs common sense as well as common rules. With diseases that spread slowly, such as some root diseases, the second rule may seem to have no urgency. But these diseases spread in time, as experience shows, and the rule applies in time.

G. The Reduction of Disease by Farm and Country Planning

What is discussed under this healing in five paragraphs, namely, the planned reduction of disease by the proper spacing and grouping of fields, may well have in the future a chapter and then a treatise to itself. This is sanitation, no different fundamentally from the destruction of alternate hosts such as barberry bushes. But we are concerned here not with other hosts, but with the crop itself. Two examples will illustrate the problem. One of the chief reasons for epidemics of virus yellows of beet in England is the indiscriminate intermingling of the sugar and fodder beets with some 8000 acres of seed crops in areas naturally congenial to the aphid vector. Concentrating seed growing into larger fields in fewer regions would reduce the chance of epidemics. In Holland 80% of the surface of the area, De Streek, is down to potatoes, mainly the early-maturing variety Duke of York, which is intensely susceptible to blight. This (presumably) has come about entirely for reasons unconnected with blight, such as the suitability of the soil and climate for early varieties. But the fact remains that the concentration of very susceptible early varieties is excellent planning of a countryside against blight, because the over-all damage is less than when very susceptible early and less susceptible mid-season varieties are intermingled.

The problem is to compute the effect of a change in the spacing and grouping—the pattern—of fields. The simplest way is to put matters on a relative basis, to use the existing pattern as a standard, and to compute how a change in the pattern would change the date of the onset of an epidemic. From the gradients one can determine the percentage change in the number, i.e., the relative number, of daughter lesions that a field acquires from other fields. From the percentage change one can use equation 11 to determine Δt, the delay in the onset of the epidemic, i.e., the relative date. If, to quote for illustration purely random figures, Δt is 6 days, the date on onset might be postponed from July 20 to 26, or from August 1 to 7, or from November 12 to 18, the actual date, as distinct from the delay, will not be estimated in advance. But the estimate of the delay itself is enough for an estimate of the gain in yield (from a knowledge of average dates of onset and average increments of yield with the development of the crop). From this one could judge whether any feasible change in the pattern of fields is worth recommending.

It is not implied that the calculations would be simple, but ways and means of numerical computation can usually be found. The determination of gradients, especially over distances not near the source, may be difficult; among fungi, a start might be made with those that release their spores only in the cool of the night and early morning or during cloudy, wet weather; their gradients are likely to be more easily determined. But collecting an appropriate range of r values for different varieties, different maturities, and different conditions of the weather should not be difficult; for some diseases such as potato blight, data already exist in the literature, form which a range of values could be calculated. The response of the various diseases will vary. The greatest benefits from a change in the pattern of fields—often a delay of months in the onset—can be expected when the gradient is steep and r is low. (Steep gradients will also simplify the computations because they limit the area that must be scanned).

Diseases such as potato blight, with high values of r but steep gradients, are less apt for inclusion in planning but cannot be excluded.

Diseases with both shallow gradients and high values of r are probably not worth planning for. This lack of universality does not condemn the method any more than the use of fungicides in plant pathology is condemned by their failure (up until now) to help much in controlling the rust diseases of cereals. One chooses a method by what it will do, not by what it does not do. The general scope for planning is becoming more favorable under both poles of farming, capitalistic and communistic, because mechanization is introducing larger units are more freedom for organization. On the evidence, there is already room for maneuver, at least with some diseases.

H. Epidemics in Experimental Plots; an Epidemiological View of Field Experiments in Plant Pathology

In Section II, E we discussed secondary epidemics. Potato varieties, which although not immune are resistant enough to escape epidemics on their own, can become severely infected through the multiplication of inoculum received from susceptible varieties. Apples of the variety Delicious have resistance but not immunity to powdery mildew, and may become infected when interplanted with susceptible Jonathan as pollinators. So, too, when one plants small experimental plots of resistant but not immune varieties in the same variety trial as plots of susceptible varieties, one can expect inoculum to move from the susceptible to the resistant varieties and initiate there a secondary epidemic that otherwise would not have developed. If the purpose of the trial is a demonstration of qualitative differences, no harm is done. But if the purpose is to put a quantitative value on the resistance, this value will be underestimated, possibly grossly, and the trial will give no indication of how the variety would behave in the hands of the farmer, which after all is the point of the trial.

Smallness of plot magnifies the error because it magnifies movement of inoculum between plots. Also, resistance itself contributes to the error: the greater the resistance, the smaller the value of r, so, by equation 11, the greater the advancement of the date of onset of the epidemic in consequence of the arrival of a given amount of inoculum from the susceptible varieties, and hence the greater the error in evaluating resistance, if the onset is early enough to affect the result. The first rule of sanitation (Section II, C) prevails in experimental plots as much as elsewhere; here it is in reverse, with an advancement instead of a delay of the onset (i.e., with Δt negative). Precisely the same argument holds in trials with fungicides. If the fungicide were perfect and conveyed complete protection in all circumstances, no harm would be done by the proximity of unsprayed control plots. But fungicides ordinarily fall short of perfection, and the presence of unsprayed controls necessarily causes the value of the fungicide to the farmer to be underestimated. An example has been given by Christ (1957). A fault on the right side, one may say, but it remains a fault.

There has been a strange change of fashion. In the bad old days experimenters dispensed with adequate replication. Their experiments were often biologically sound, but seldom capable of statistical interpretation. Nowadays we determine with great mathematical nicety the statistical significance of a result that has no reality outside the experimenter's plots. It seems that the statistician and the plant pathologist have each wrongly assumed that the other has examined the techniques of field experiments in plant pathology and approved of them. Better techniques could probably be evolved, perhaps by using plots of unequal size and cutting down subdivisions as much as is possible without interfering greatly with statistical efficiency.

But the real solution is to reduce the errors at their source. If, for demonstration purposes, it is necessary to plant a resistant and susceptible variety together, or to have a sprayed and unsprayed plot together, it should be considered as a demonstration and left at that. But if one wishes to evaluate new varieties quantitatively, they should be compared with others of about the same class: susceptible varieties with susceptible varieties, moderately resistant varieties with moderately resistant varieties—each class in its own separate trial. In fungicide trials one fungicide should be compared with another without an untreated control, if the purpose of the trial is to determine which fungicide is the more efficient. This will reduce the error, even if it does not entirely exclude it. From these remarks one excepts diseases like many root diseases that spread too slowly for plots to interfere with each other.

V. The Host Plants

A. Conditions for an Epidemic of Disease in Annual Crops

1. *The Annual Rise and Decline of Epidemics: Rules about its Form.* Among annuals we include perennials and biennials growth as annuals, such as beet growth for its growth for its roots. Epidemiologically, the problem with annuals is to determine how disease can strike quickly enough to cause substantial damage. Heavy losses can occur quickly enough if small lesions multiply fast enough to compensate for scarcity of inoculum at the start, or if the amount of

inoculum at the start is high enough to compensate for a slow rate of multiplication of small lesions, or if each individual lesion causes damage great enough to compensate for low initial inoculum or slow multiplication, or with any of the countless gradations between these ways.

Consider the first two ways. According to equation 11 the higher the value of *r*, the less the effect on an annual epidemic of variations in the carry-over of inoculum from the previous season, and hence the less the effect of environmental factors other than those operating during the season of the epidemic itself. The climatic and other risks are concentrated and not well spread, and, as a general rule, one can expect large fluctuations from season to season. The same rule can be put more elaborately: the higher the value of *r*, the less likely is a nonrandom succession of epidemic years, provided that the variety of the host and the race of the pathogen remain unchanged. The provision is necessary to exclude what happens when a new aggressive race arises and causes successive epidemics until it is countered by new resistant varieties.

Stem rust of wheat, caused by *Puccinia graminis*, illustrates this rule. The explosiveness of its epidemics indicates a high value of *r*, and the data compiled by McCallan (1946) show a wide fluctuation—from a trace to 23%—in crop losses from year to year in the United States. Leaf rust, caused by *P. triticina*, which on the whole produces less explosive epidemics, does almost as much damage on the average, but within narrower limits, from 1.0 to 9.6% as one would expect from a lower value of *r*. Another rule, which needs little explanation, is: the higher the value of *r*, the steeper the rise, and, usually, the steeper the decline of an annual epidemic. The second part, about the decline, can only be inferred logically for low-priced crops such as corn and wheat. With high-priced crops that justify a large fungicide account one cannot infer that inoculum drops to a low level between seasons, even if *r* is high.

2. *The Fast Multiplication of Small Lesions with a Low Amount of Inoculum at the Start*. Because of a high value of *r*, sanitation is not very effective against diseases with these characteristics (Section II, C) unless it is undertaken thoroughly and extensively, as in a nationwide eradication of barberry bushes for the control of stem rust of wheat. Epidemics are typically controlled by using resistant varieties (e.g., against late blight of tomato). Epidemics are typically widespread; they involve more than a few farms and commonly extend over whole

countries of states. This accords with what was said in Section III. Epidemics tend to fluctuate widely from season to season.

3. *The Slow Multiplication of Small Lesions with a High Amount of Inoculum at the Start*. The root knot nematodes (*Meloidogyne* spp.) of tropical and subtropical climates are examples. They survive well from year to year. The multiplication rate is relatively low; e.g., one does not expect severe epidemics if the soil is only lightly infected at the beginning of the season. This is indeed the fundamental assumption in soil fumigation, which reduces the population of nematodes but normally falls far short of complete eradication. Because of a low value of r, diseases in the category can usually be controlled by sanitation, e.g., by fumigation, isolation, clean fallows, or crop rotation. Epidemics are typically local, i.e., they may involve only a part of a single field. Fluctuation from year to year are relatively small; one can normally predict fairly accurately what will happen one year from what happened the previous year.

4. *Multiplication of Large Destructive Lesions and Systemic Infections*. If each individual lesion or systemic infection is very destructive, relatively slow multiplication of a relatively small amount of inoculum at the start will bring about a destructive epidemic. Because of slow multiplication, sanitation is commonly effective (Section II, C). Thus, for virus diseases of annuals, sanitation is the commonest recommendation for control.

An analysis of recommendations in the 1957 edition of Smith's "Textbook of Plant Virus Diseases" establishes this point. Epidemics tend to be local (Section III, G). Evidence here is strong; there does not seem to be a single example among annual plants of a disease with large lesions that spreads widely in a year without man's help. One expects disease not to be fluctuate wildly from year to year. Evidence here is weak and conflicting. Hull (1952) states that epidemics of virus yellows in sugar beet develop in a spiral over a period of years. Peak outbreaks have never yet occurred in the root crop in England directly after years of light infection, but have taken 2 or 3 years to develop. But Wolf (1935) found no correlation between the amounts of mosaic in consecutive tobacco crops on the same land, even though in 125 out of 229 fields the virus was observed to overwinter in tobacco stubbles.

B. Conditions for an Epidemic in the Annual New Growth of Perennial Crops

There is the same need for speed here; the new growth must be infected quickly enough for the epidemic to be destructive. Some diseases owe little to the perennial nature of the host; e.g., tuber-borne inoculum does not usually play a significant part in epidemics of early blight in potatoes, caused by *Alternaria solani*. The greatest effect of perenniality occurs when the crop occupies a perennial site such as an apple orchard or vineyard or when the disease affects perennial tissues as well as the annual new growth. Innoculum then commonly overwinters locally; e.g., epidemics of apple scab start from inoculum released in spring from rotting leaves or infected twigs; epidemics of apple and grape powdery mildew start from infected bud scales.

Many of the diseases of the annual new growth of perennial crops are caused by fungi that multiply fast in small lesions. They differ from corresponding diseases of annual crops in often having a larger initial source of overwintered inoculum. Consequently many tend to be particularly destructive, and must be kept under control by the lavish use of fungicides. It is a matter of fact that the greatest sales of fungicides (excluding seed dressings) are for use against diseases in this category. The crops must be productive enough in terms of money to pay the bill.

C. Epidemics of Disease in Perennial Tissues

1. *Plants Grown from Seed*. The diseases in coniferous forests fall within this group, as well as many diseases of hardwoods, plantation crops, ornamentals, and the like. In agriculture one is concerned largely with epidemic disease; in forestry, largely with endemic disease. With indigenous pathogens of indigenous trees growing under natural conditions, inoculum has existed over a long period, and a balance is struck between pathogen and host. Disease control is primarily a matter of: *a*, forestry practice to maintain a reasonably healthy balance; *b*, selecting suitable sites for species adapted to the locality with stok grown from seed of suitable origin and with the appropriate composition and management of the stand.

There may be small local and temporary epidemics, but the general ecological pattern is of a stable community of pathogens and hosts. In the language of tern is of a stable community of pathogens and hosts. In the language of this chapter there is no multiplication and no spread; inoculum is perennial, and within limits the area of the forest has little effect. What we have been discussing is inapplicable—which emphasizes the point that this chapter is concerned only with epidemic disease and not disease in general. With introduced pathogens or introduced hosts a new balance must be struck, and it is commonly very unfavorable to the host. Examples are chestnut blight, Dutch elm disease, swollen shoot of cacao. Epidemics of disease following new combinations of host and pathogen seem to be much like those of disease in annual plants, except that the time scale is different; one can conveniently plot the progress of the epidemic in years rather than weeks.

2. *Dangers of Vegetative Propagation.* (a) *Special Danger from Systemic Disease.* Although the material used for vegetative propagation can carry to inoculum of local lesions, it is especially dangerous for systemic disease. In "Plant Diseases", the 1953 Yearbook of Agriculture of the United States Department of Agriculture, about 8% of the space given to diseases of plants grown from seed is about viruses. For vegetatively propagated plants the figure is 28%. It is largely because of the property of systemic infection that viruses are so dangerous in plants propagated vegetatively.

(b) *Danger from Longevity.* Many of the clones used in agriculture and horticulture are old. The bulk of the citrus fruit of commerce comes from varieties 80 years old or more. Old apple varieties remain highly popular. The potato varieties Russet Burbank, Irish Cobbler, and White Rose, which rank second, fourth and sixth, respectively, in the United States, are all very old. This longevity provides the span needed for a slow epidemic—what in Section I, B was called a low death rate epidemic. It is one of several factors in the accumulation of virus diseases, so that, for example, it is rare to find a single healthy clonal citrus tree, as experience with the Florida budwood certification scheme shows. The longevity factor (the danger of perpetuating the virus) is the one usually considered in references to the danger of virus diseases in vegetatively propagated plants. But it is easy to overrate the importance of the factor. In providing a long span of years long-lived clones are no different from long-lived plants grown from seed.

(c) *Danger from the Randomization of Sources of Infection.* Figure 2B shows a nest of leaf-roll-infected potato plants loosely clustered about the plant that was the original source of inoculum. Within the nest, and especially along the rows which are the main direction of spread, there are many nonrandom contacts between diseased plants: contacts that are harmless in the spread of a systemic disease. Within the nest diseased plants are separated from healthy plants outside the nest; the deeper they are within the nest, the greater the separation and the less the chance of transmission. Suppose that the field of potatoes had been harvested and the crop used for seed the next year.

There would have been a mixing of seed with a scattering of diseased tubers among healthy ones, so that when the new crop grew there would be new random contacts between diseased and healthy plants that would allow a less hindered spread of infection. The phenomenon is general.

When systemic disease spreads from a focus in an orchard to form a nest of diseased trees, there is a similar automatic brake on further spread and a similar release of that brake every time buds are randomly collected as propagating material for a new orchard. The danger from this source is greatest when r or r' is small, as it usually is with systemic disease of trees in particular. For reasons given in discussing scales of distance in Section III, E the smaller the value of r', the greater the tendency for diseased plants to cluster, hence the stronger the automatic brake and the greater the benefit to the pathogen of a random use of propagating material in establishing a new orchard or field. One could paraphrase this in general terms by saying that the feebler the pathogen's own power of spread, the greater is the relative benefit to it of man's moving propagating material around, and that feeble powers are on the whole likely to be found in systemic pathogens of trees.

(d) *Danger from the Infectious Incompatability of Scion and Rootstock*. One of the worst epidemics on record, that of tristeza disease of citrus in South America, resulted from what in the event proved to have been an unfortunate choice of rootstock. Typically, tristeza was found in sweet orange on sour orange rootstocks although other combinations were involved. Neither sweet orange on sweet orange stock nor sour orange on sour orange stock suffered noticeably. The disease was one of a combination of species, not of single species. The key to the understanding of tristeza is the difference between transmission by grafting and by vectors. Infected sweet oranges harbor with apparent tolerance a virus component which is readily transmissible to sour orange seedlings by grafting, and causes them to be stunted. But although this component is freely transmitted by vectors such as *Toxoptera citricidus* from sweet orange to sweet orange, it has not been found in adult sour orange trees even when they grow beside infected sweet oranges in a vector-ridden orchard. Sour orange in the orchard is resistant to systemic infection by vectors of this particular component.

But when one grows a sweet orange scion on a sour orange rootstock in a tristeza-infected locality, the sweet orange foliage acquires the component by vector transmission, and the sour orange rootstock acquires it from the scion across the graft union. The result, it seems, is tristeza. To generalize, the necessary conditions for infectious incompatibility between species neither of which shows marked symptoms on its own roots are probably that both species should be susceptible by graft transmission, that one or both species should be resistant to systemic infection by vectors, and that one but not both species should be tolerant of the pathogen or particular strain of it. Exocortis of sweet orange on trifoliate rootstocks is probably an example in which both species are resistant to systemic infection by vectors (no vector is known). Incompatibility in such a combination is normally the result of using infected material in the nursery.

(e) *Pernicious Nursery Practices; Contamination During Handling.* It would be unrealistic to ignore the part that pernicious nursery practices can play in increasing diseases which on their own increase slowly. Rootstocks have been rebudded after the first buds have failed, sometimes with a different variety. Rose nurserymen have been known to obtain an abundant source of material for rootstocks by cutting stocks off above the bud union and reusing the cuttings after rooting them for a new lot of buds the following season. Rather different in type because the practices are not in themselves reprehensible is the spread of inoculum while handling the material for propagation. Ring rot, caused by *Corynebacterium sepedonicum*, black leg, caused by *Erwinia phytophthora*, and the virus disease spindle tuber are transmitted by the knife used to cut seed potatoes, and, other things being equal, epidemics are more common when seed is cut and not planted whole.

(f) *Genetic Uniformity of Clones.* Apart from mutations, a clone remains genetically uniform. If a pathogen can attack the clone, genetic conditions are uniformly favorable to disease—a point sometimes stressed. On the other hand, if the clone is resistant, conditions are uniformly unfavorable to disease—a point sometimes overlooked. The problem is part of a wider problem discussed in Section V, E. The danger of secondary epidemics in crops not uniform in resistance has been mentioned in Section II, E.

(g) *Disease in Relation to Vegetative Propagation in Nature.* Plants propagated vegetatively are common in nature, ranging from sod grasses to suckering trees. As a group they do not seem to be especially liable to disease. In them vegetative propagation is primarily a matter of longevity and (presumably) of genetic uniformity. In particular, there is no randomization of sources of infection. Members of the clone stay close together, even when propagation material is released from the air, as the bulbils of *Agave*. To put the matter teleologically (purely for the sake of brevity!), nature does not make the mistake of distributing this material as she distributes seeds and fruits. Only man makes this mistake, and gives the pathogen a mobility it would otherwise not have. The dangers of vegetative propagation usually stressed in the literature are the dangers of perpetuating the pathogen in the clone, i.e., the longevity factor, and the danger of genetic uniformity. If those were the worst dangers, the prospects of improvement would not hopeful, because the dangers would be inherent in the core of the process of vegetative propagation. But with vegetative propagation in nature as a background one doubts whether they are. An assessment of disease factors in vegetative propagation in agriculture and horticulture seems overdue.

In particular, it should be assessed whether vegetative propagation is in fact inherently dangerous or whether we are not making it more dangerous than it need be. If, as one might guess, much of the trouble starts from the randomization of sources of infection, it should not be beyond the ability of man to devise means of curbing the randomness. Partial curbs are already being applied: perhaps the best-known is tuber-uniting of potatoes (cutting seed potatoes and planting the pieces together in sequence). But more could be done, and pathologists might take a broad look at the disease problem in vegetative propagation to see whether they cannot cut it down to size.

D. Epidemics of Systemic Disease

1. *Relation with the Size of the Plant*. In Section III, F a systemically infected plant was regarded as a single large lesion. Between small plants growing close together in narrow rows inoculum has less distance to travel than between large trees proportionately widely spaced. Other things being equal, an individual vector, for example, will transmit disease more easily between the small plants than between the trees. But there in another cuite different problem. If large plants are scattered among the small plants without change of spacing, the large plants are a bigger target for the inoculum and are more likely to become infected. Because it is implicit in the concept of systemic disease that only one successful transmission is needed to infect the plant, these large plants will be expected to develop a larger percentage of systemic disease than the small plants among which they are scattered. Some figures of Broadbent (1957) can be quoted. In cauliflower seedbeds the infection will cauliflower mosaic

was 37.5% in large plants, 6.5% in medium plants, and 0.5% in small plants. The corresponding figures for cabbage black ring spot were 13.8, 7.0, and 2.4% respectively.

2. *Relation with the Number of Plants per Acre.* Suppose plants are being infected with a systemic disease caused by inoculum coming from outside the plants, e.g., plants infected with air-borne or vector-borne inoculum entering the field from another field or plant infected with inoculum from the soil in which they grow. Consider, for example, two fields exposed to a uniform invasion of migrating infective insects. If m_1 and m_2 are the mean number of transmissions of inoculum per plant in the two fields an d x_1 and x_2 the proportions of disease developed,

$$1 - x_1 = e^{-m1}$$
$$1 - x_2 = e^{-m2}$$

according to the Poisson theorem. Hence,.

$$\frac{\log(1-x_1)}{\log(1-x_2)} = \frac{m_1}{m_2}$$

The mean number of transmissions per plant can reasonably be assumed to be proportional to the number of vectors per plant, and, because the vector environment of the two fields is uniform, this is inversely proportional to the number of plants n_1 and n_2 per unit area in the two fields. Thus,

$$\frac{m_1}{m_2} = \frac{n_2}{n_1}$$

and

$$\frac{\log(1-x_1)}{\log(1-x_2)} = \frac{n_2}{n_1} \qquad (15)$$

This relation (in a slightly different outward guise) was used by van der Plank and Anderssen (1945) for the discussion of infection of tobacco with tomato spotted wilt virus. The virus is brought into fields by thrips which apparently settle randomly and do not spread the disease from tobacco to tobacco. The percentage of disease varied with the number of plants per unit area in the manner predicted by equation 15.

If the proportion of infection is low for a given disease and vector, one may simplify matters by an algebraic transformation, and, as an approximation, write

$$\frac{x_1}{x_2} = \frac{n_2}{n_1}$$

If μ_1 and μ_2 are the number of infected plants per unit area in the two fields, then

$$\mu_1 = n_1 x_1$$
$$\mu_2 = n_2 x_2$$

So, as an approximation when the proportion of infection is low.

$$\mu_1 = \mu_2. \quad (16)$$

With systemic disease, if inoculum enters the crop randomly and if the proportion of infection is low, the number of infected plants per unit area is constant and independent of the density of the stand. This relation was observed experimentally by Linford (1943) without a realization of the condition that the proportion of infection must be low. Pineapple plants were spaced 12, 15, and 18 inches apart in the row, giving populations of 21, 780, 18,150 and 14,250 plants per acre, respectively. The same number of plants per acre developed yellow spot, caused probably by tomato spotted wilt virus and carried in randomly by thrips.

The percentage of infection was low (3.3, 4.6, and 5.2% for the 12, 15, and 18 inches spacing, respectively), so the condition underlying equation 16 is satisfied, and the number of infected plants per acre was 720, 820, and 760, respectively. In modern agronomy there is a tend toward high plant populations per acre. From the point of view of disease levels this is a trend in the right direction for crops menaced by vascular wilt diseases and other systemic or quasi-systemic diseases. A simple experiment makes a useful demonstration: tomato seedlings growing sparsely spaced in trays are easily infected at suitable temperatures with bacterial wilt when a culture of *Pseudomonas solanacearum* is poured over the soil, but it is much more difficult to infect a high proportion of tightly crowded seedlings.

Forecasting Epidemics

Man found his crops destroyed by rusts and smuts, mildews and blights long before he recognized the microscopic pathogens. Some men correlated disease with sinfulness; other, finding only a slight variation in the amount of sin and a large variation in the weather, obtained a better correlation—that between weather and disease. Consequently, they called the weather the cause of plant disease. In this imperfect state of knowledge they were able to forecast disease outbreaks before they were able to name the pathogenic fungi. Now, in our still imperfect state of knowledge, we too can forecast disease outbreaks from our science of the interaction of host, pathogen, and weather. The forecasting of epidemics is a contribution to the forecasting of crop yields. Yield forecasts are useful if they arrive sufficiently early to permit adjustments in acreage or transportation or if remedies can be applied to prevent the prophesied disaster. An empirical relation between weather and yields can assist in adjustments. The relation becomes more logical and satisfying and is productive of remedies if it also includes a knowledge of the contributions of soil fertility and of varieties to yield.

Of primary interest here, the relation is more logical and fruitful if it includes a knowledge of the interaction of pathogens with weather, soil, and host; this is most fruitful if a remedy can be applied after the forecast is made. Reviews of the forecasting of epidemics of plant disease have appeared in the past. Because the weather relations of diseases are often obvious, the practice is a venerable one, and the many examples over the world have been examined by Foister (1929) and by Miller and) O'Brien (1957). These reviews provide complete coverage. Therefore, the subject will not be neglected if the present examination takes a different path: the etiology of plant disease has been examined for opportunities for prognostication of disease and subsequent loss. Illustrative examples will be cited where possible.

The forecasting of epidemics is quantitative epidemiology applied with courage. The better to make it quantitative, a formal framework is proposed and employed as follows: The number of spores reaching a host will double as the production is doubled or arrivals are proportional to Q, the number of propagules released. The arrivals also increase as the trapping efficiency p increases; however, the relation is not linear because the disseminating "cloud" of pathogens is depleted by this trapping. Thus, arrivals are proportional to $pe^{-\text{constant } pf(X)}$ where $f(X)$ is a function of distance X. The number of spores per volume of air or other medium decreases rapidly with distance X or arrivals are proportional to $1/X^n$; since dissemination is generally three-dimensional, n is about 2. Finally, the number of infections D is related to arrivals or D = constant $[(S/K)\ p/X^n]\}e^{-\text{constant } pf(X)}$. The K is the proportionality constant between arrivals and infection. For the important case of aerial dissemination, estimates of the parameters have been made: p can be about 1/20, n about 2, the $f(X) = X^{1/8}$. As the etiology is examined for possibilities for prognosis, these possibilities are evaluated in terms of the above model.

I. The Primary Inoculum

Many factors will affect the magnitude of the explosion called an epidemic, but the spark that ignites, i.e., the primary inoculum, must be present. And, as the model shows, the greater the quantity Q, the greater the epidemic likely. Thus, the quantity of primary inoculum can be a criterion for forecasting, a criterion particularly useful because of its early appearance. The diseases of crops grown in the temperate zone have been the objects of most forecasting schemes. Here the pathogen often must assume a resistant form, take shelter, become dormant, or be eliminated and renewed from the tropics. The forecaster can examine these survival mechanisms for his first hint of the new season's prospects. An early indication of the amount of overwintering primary inoculum is the severity of disease in the previous season. Although many factors can modify the amount, a persistence in amount is seen from year to year in certain diseases due in part to the greater primary inoculum carried over after an epidemic year. For example, the persistence of the catastrophic severity of potato later blight from year to year increased the disastrous consequences of the appearance of *Phytophthora infestans* in Ireland in the "hungry forties." The appearance of the new *Helminthosporium victoriae* led to the persistent decimation of the varieties derived from Victoria oats and necessitated the introduction of new varieties. Thus, the "grand cycle" of disease, the steady increase and decrease of severity from year to year, caused partially by the overwintering of inoculum is an early straw in the wind indicating the prospects in a new season.

If an overwintering, resistant—often perfect—stage of the pathogen is known, the procedure is more satisfying than complete dependence upon the "grand cycle." Here the forecaster can search for the overwintering form itself and see the spark that may ignite the coming explosion. Many classic schemes for forecasting apple scab proceed by this means. The pathogen *Verturia inaequalis* overwiners in the perfect stage in fallen leaves where its maturity is dependent upon the winter and early spring weather. Microscopic examination of the perithecia, ripening in the warmth of the laboratory, and considerations of winter weather have formed the bases for prediction of the first discharges of ascospores in the spring. Forecasts of infection then can be made from knowledge of the stage of tree development, the response of growers to the spray warnings, and subsequent discharge and infection periods.

Brown rot of peaches, caused by *Monilinia fructicola*, is another ascomycetous pathogen that overwinters in the perfect stage although the imperfect stage is also functional. A search of orchards for the cankers and mummified fruit that harbor the par reveals primary inoculum for the succeeding season, and could form the basis of a forecast. Ergot of cereals, especially rye, is another classic disease that overwinters as an easily visible, perfect stage. The discovery of a multitude of the sclerotia of *Claviceps purpurea* in seed or even in a field can be a warning of an epidemic in the field to be planted or in the field where sclerotia may have fallen. A pathogen may take shelter during the cold season, and a forecasting scheme can be founded upon the amount sheltered. For example, *Bacterium stewartii*, the pathogen of bacterial wilt of sweet corn, survives the winter in the bodies of adult flea beetles. The survival of the sheltering beetles is encouraged by warm winters.

Stevens (1934) found that severe epidemics of wilt followed warm winters, and, hence, founded a forecasting scheme employed successfully by him and Boewe in Illinois and by other workers in New Jersey and New York. A pathogen can invaded seed and be sheltered for the winter with the crop. *Phytophthora infestans* infects tubers in the autumn and overwinters in the storage bin. Here the inspector can estimate the amount of primary inoculum available for the following season. Wallin (1956) has done this with some success. Of equal importance when blighted tubers are culled from the storage and dumped, the sprouts on the cull piles are a fruitful source of inoculum and an opportunity for the forecaster to take a census of primary inoculum.

The extent of virus infection can be accurately predicted by indexing potato seed, a practice generally followed. Loose smut of wheat or barley derives from seed infected by *Ustilago tritici* or *Ustilago nuda* during the previous season. Thus the census of the infection in the seed can lead to a forecast of epidemics; the detection of infection can be made by germinating or by examining microscopically a sample of seed. A forecast can also be based upon an even earlier event, the weather at blossoming times in the preceding season, because this is the time of susceptibility. Some pathogens lie dormant awaiting the arrival of warm weather and a host. Their numbers are an index of epidemics to come. This is the case of many soil-borne pathogens—pathogens that spread relatively slowly. The extent of *Phymatotrichum omnivorum* and its depredations can be predicted almost to the foot from its extent in the previous year. Nematode infestation can be forecast because of the persistence of the worms or cysts. Whether or not a crop will be infected by *Verticillium alboatrum* has been predicted from the response of an index plant, the tomato, grown in soil samples from the intended field. If rotation decreases the population of a pathogen, severity of infection can be predicted from the intensity of management.

Pathogens can lie dormant in an overwintering crop. If the crop, like winter wheat, is harvested in midsummer, the pathogen must race against time if an epidemic is to occur. Hence, the overwintering fungus and its early multiplication are critical. Thus, Chester (1946) found that *Puccinia rubigo-vera* overwinters in Oklahoma in pustules of wheat leaf rust; the level of infection on April 1 is determined by the weather in the late winter; and this level of infection determines the extent of any epidemic, because weather after that date is rarely limiting. Forecasts by this method have been eminently successful, the only failure bring caused by the rare event of later weather limiting infection. A pathogen may be largely eradicated from a region by the winter, and primary inoculum may be borne in on southerly winds. Following the frequent observation of fungal spores in the upper atmosphere, much attention has been devoted to this concept. For example, the annual renewal of *Puccinia graminis var. tritici* and the production of epidemics of stem rust of wheat has been attributed after extensive studies by Stakman and co-workers to spores borne aloft in Mexico and showered down upon the Great Plains. Epidemics of tobacco blue mold in Quebec have been associated with the production of *Peronospora tabacina* in Kentucky and a forecasting scheme devised accordingly. Two considerations should be kept in mild in evaluating the foregoing.

First, as spring progresses northward, even diseases of local origin will appear to move northward. Second, the rapid dilution by distance of spore clouds, as evidenced by the steep gradients of infection about isolated sources makes the probability extremely small that spores from continental distances will alight on a given field. The second consideration renders spore traps of little use in forecasting; Rusakov found no spores upon traps until 1% of the plants in the neighborhood had been infected, and he suggested a local field inspection instead of dependence upon traps. The characteristics of stem rust make it more susceptible than many diseases to a forecast based upon spore movement from the South. The astronomical number of spores produced in the large acreages of wheat in the South and the large acreages of hosts to the North somewhat counteract the dilution by distance.

The sturdiness of the *Puccinias* spores assures that they will be viable when they arrive. The broad and nearly continuous belt of wheat extending from Texas to Saskatchewan permits *Puccinia* to leap, frog-like from country to country, never requiring it to move continental distances at a single jump. Thus, a forecasting scheme for stem rust can logically be based upon the systemic movement of primary inoculum from the South. The sensitive spores of the downy mildews, produced on limited acreages and seeking out limited acreages separated by miles deserted of hosts, presents a contrasting picture. Thus, a survey of the inoculum available at the beginning of a new season provides a rational point of departure for a disease forecast. It has the advantage of earliness, with the consequent opportunities for remedial measures, such as selecting an alternate crop, roguing, or chemical control. It has the disadvantage of occurring so early that many factors can subsequently modify the outcome. Nevertheless, the forecaster courts disaster if he does not issue a word of warning when the inoculum potential is obviously high, or, on the other hand, if he forecasts an epidemic solely from weather data when the pathogen is absent.

II. Dispersal of Inoculum

When a given amount of primary infection is present, the successful pathogen must next produce units such as cells or spores, and have them detached. Garrett calls these units propagules. The product of the primary infection times the output of propagules per lesion comprises the source strength of *Q* and the subsequent epidemic is proportional to *Q*. Hence, another opportunity for forecasting is presented. Here, in considering the production and detachment of propagules such as spores, wheather is met in all of its effectiveness.

Many diseases, especially mildews, have long been associated with damp weather; the late blight forecasts of Martin (1923) and Cook (1949) were based upon average or accumulated rainfall and a average temperatures. This is, however, an oversimplification. Crosier (1934), interpreting his own and Melhus's studies of the biology of *Phytophthora infestans*, wrote, "It is not the total rainfall nor the average temperature, but the coexistence of moisture and low temperature (from 10 to 15^{0}C.) for the one-half hour or longer, that makes possible the formation of swarmspores, and ... infection follows promptly if the moisture persists." Further, he emphasized, "No swarmspores can be formed, irrespective of the temperature and moisture, unless viable sporangia are present." The present discussion, following the life cycles of pathogens, attempts to recognize these fundamentals. Crosier, in the above study, found sporulation was most abundant in an atmosphere saturated with water and at a temperature of 18 to 22^0; spornagia were abundant within 8 hours. Production was slower at lower temperatures, and sterile at temperatures 3 to 5^0 higher. Here is the necessary information for a description of an environment suitable for the multiplication of *Phytophthora* and the forecast ot late blight epidemics.

Quantitative estimates of sporualtion of this important pathogen have been made, permitting a more exact weighting of the forecast according to the production. At 21^0 sporangia appeared in 6 hours, while 8 hours were required at 18^0. If more time was permitted, most isolates produced more sporangia at 18 than at 21^0. For example, a typical isolate produced about equal numbers of spores, N, in 6 hours at 21^0 and 8 hours at 18^0; in 12 hours at 18^0 it hand produced about 20 N while at 21^0 it has produced only 3 N. The forecaster should beware of environmental races, however; one isolate was able to produce between 1 and $2N$ sporangia in 12 hours at the high temperature of 27^0. Employing data of this type and a survey of primary infection one can estimate the production of spores from hygrothermograph records, and a forecast can be begun.

The final step in the estimation of Q, the number of propagules dispersed, is an examination of the discharge of the units into the air. The units have been attached to the host for nutrition; now they must break this attachment, pass through the enveloping 100 microns of the laminar flow layer of air, and reach the turbulent air above. Many viruses are borne by insects, and their numbers at critical times in the host's life can form the basis of a forecast of epidemics. Some bacteria are borne in splattering raindrops; the arrival of rain might be employed in forecasting a bacterial epidemic.

Fungi have developed many clever devices for dispersing their spores—devices that have been studied by deBary, Buller, and Ingold. Among these are some that depend upon the weather and,.hence, can be limiting and of interest to the forecaster. Tobacco blue mold is a case in point. DeBary observed how the sporangia of some Oomycetes are discharged as the sporangiophores dry and twirl; Pinckard (1942) found this phenomenon specifically in *Peronospora tabacina*. The disease tobacco blue mold does not increase during prolonged rains; spores are found in the air as dew or rain dry, not while the lesions are wet. From these observations the forecaster of blue mold can learn to look beyond rains and dews and consider the critical hours after the leaves have dried and the spores have flown. This completes the estimation of the number of spores released, the source strength Q to which the subsequent epidemic will be proportional if other things are equal.

This early omen depends upon the quantity of primary inoculum and infection and upon conditions favorable for the production and release of the propagules. The omen appears to be forecaster early enough for chemical control to be limited, probably not early enough for the selection of an alternate crop or for roguing. Many hurdles must yet be taken by the pathogen before an epidemic occurs, but the presence or absence of spore production is a valuable, early clue.

III. The Transfer of Inoculum

The propagules, once separated from its parent, can be carried to an infection court on some object, in the soil or in water, or through the ar. This transfer is generally not under the control of the pathogen, can be limiting, and, hence, is an opportunity for forecasting. Many contaminated objects such as seed, soil, and machinery are borne by man. The speed and volume of commerce, the wide distribution of hosts—particularly economic species—and the prolific nature of pathogens causes the probability of permanent nonintroduction of new pests into a region to approach zero. A knowledge of sanitation and of quarantine measures, as well as the foregoing factors, permits the forecaster to estimate how soon the inevitable introduction and subsequent epidemic will come.

The annual reintroduction of pests can be forecasts in the same way. For example, the coming incidence of potato ring rot, caused by *Corynebacterium sepedonicum*, can be anticipated from the liveliness of its transfer. Is seed inspected? Is contaminated equipment sterilized? Is whole seed used? If not, then an epidemic can be safely forecast whenever the inoculum was introduced into the seed-growing regions during the previous season. Many viruses are transferred in insects, particularly aphids. As the number of aphids Is decreased, transfer—as is dispersal—is decreased. This may be due to a northern climate, e.g., that of Maine or Scotland; the time of the year; an exposed site; or sometimes an insecticide. Whatever drastically reduces the population of insect vectors reduces Q, and a lower incidence of infection can be forecast.

In many cases a sharp decrease in infection is observed at increasing distances from the source of aphid-borne viruses. In another case the sharp decrease observed by Waggoner and Kring (1956) became a broad distribution when the susceptible Green Mountain variety and extremely high insect populations were encountered. Nevertheless, under the usual field conditions of moderate populations of insects and some host resistance, the decrease with distance will be sharp, disease being proportional to $1/X^n$ where n is about 2. This is not surprising; Wolfenbarger (1946) has catalogued many insect dispersal patterns and found them similar. From this the forecaster can derive two useful conclusions: disease severity will be slight at considerable distances from sources, especially in a single generation of the pathogen; and numerous widely scattered sources are necessary for an epidemic.

The spread of insect-borne fungi behaves in a manner similar to the spread of the viruses, providing similar bases for forecasting. For example, infection by *Ceratostomella ulmi*, the cause of Dutch elm disease, increases with the population of bark beetles (which can be limited by insecticides) and with propinquity to diseased trees. Therefore, a forecast could be based upon a knowledge of the number of beetles and diseased elms in the neighborhood. In undisturbed soil the spread of pathogens can take place through the contact or grafting of roots.

The spread of *Phymatotrichum onmivorum* in cotton fields and of *Endoconidiophora fagacearum* in oak forests can occur in this fashion. Because the span of root systems is relatively small, the forecaster should be able to draw an orderly prognostic map of these diseases and see it verified.

Plant pathogens can be carried by water: rain, irrigation, or spray. If these are the channels by which a pathogen travels, its distribution can be predicted to be localized. Faulwetter's (1917) classic study of the dissemination of bacteria by splashing rain demonstrated the short range of this type of distribution. Spores suspended in some fungicide or insecticide mixtures can infect plants; unpublished experiments with *Phytophthora infestans* demonstrated that spray blasts do not increase its spread consequentially. Pathogens that spread in water evidently spread slowly.

Air-borne fungi have been classic subjects for studies in epidemiology as well as forecasting. Basing his analysis upon the statistical theory of the turbulent transfer of matter through air, Gregory (1945) has proposed a hypothesis for the transfer of inoculum; with slight modification this is the relation introduced in an early paragraph of this review. Gregory demonstrated the wide applicability of the hypothesis, infection generally being proportional to $1/X^n$, where X is the distance from the source of inoculum and n is about 2. Thus, one of the strongest clues available to the forecaster of these diseases is the nearness of inoculum: at half the distance, the probability of infection is fourfold. A concrete example can be seen in the map of an epidemic of tobacco blue mold: a few yards from the source of inoculum in an infected seedbed more than 100 lesions were present on each plant; at the far end of the 3-acre crop less than 5 lesions were present per plant; on a neighboring farm only 1 plant in 4 was infected. With the location of the source of inoculum in this neighborhood known, the course of the epidemic could be predicted for given weather and control practices.

A significant modification can be made in the average decrease with distance predicted by the inverse square relation. The spread of the pathogen downwind from the source is more rapid. Naturally, downwind refers to the direction of the wind during dissemination. Thus, the highest rate of infection would be forecast to the west of a pathogen that spread during rains. The distribution of insect-, soil-, air-borne pathogens all being commonly characterized by sharp decreases with distance from the source, van der Plank's (1949) analysis of the relative danger from field size is generally applicable. Thus, small fields and evenly distributed sources call for a forecast of a greater epidemic than do large and widely separated fields.

Two clues for forecasting have been pre-eminent in the discussion of transfer: the increased danger due to more vectors and that due to greater proximity to a source. These factors would have slight importance if the disease were permitted to run its course, but this it is rarely permitted among annual crops. Rather, the host and the pathogen race to the end of the season, and a winner is declared without a steady or equilibrium state ever becoming established. Thus, the number of vectors and the proximity of inoculum are important bases for the prediction of the damage sustained when the crop is harvested.

IV. The Trapping of Propagules

As the peripatetic plant pathogens pass over a unit area of the field, a proportion p becomes attached to the soil and to plants. An increase in this attached proportion leads to an increase in the number of infections D. Increased trapping exerts another influence: it depletes the cloud of pathogens at a more rapid rate. Thus, D is proportional to $pe^{-constant\ pf(X)}$, and a knowledge of the rate of trapping p should assist in forecasting the subsequent distribution of infection. Unfortunately, little is known about the contribution of this factor. Logical conclusions can be drawn from the above hypothesis to supplement our meager knowledge. Contrast the case of a potato infected with Y virus situated among other potatoes attractive to *Myzuz persicae* with the case of a source situated among plants unattractive to the aphid. In the first case, the danger to the adjacent plants is great, but the probability that this nonpersistent virus will be carried afar is slight; in the second case, the danger to adjacent plants is decreased at the expense of more distant ones. Now let us consider the spores of fungi—objects carried at the mercy of the winds. They vary in density, volume, and consequent settling velocity in the viscous air; presumably this might alter the proportion attached, but experience has shown the effect is inconsequential.

The ability of the spores to attach themselves to objects does, however, increase as spore size increases, leaf width decreases, or velocity increases; the small or slow-moving spores drift around a broad leaf. From these considerations more infection nearby, less at a distance, has been forecast for large-spored pathogens in tall grass; and more infection afar, less nearby, for small-spored pathogens in low, broad-leaved plants. Verification is lacking but the logic is compelling. More than a forecast of distribution can be assisted by the foregoing. We have seen in preceding sections the necessity of a nearby source of propagules if infection and losses are to be consequential at harvest. From this it follows that widespread sources and subsequent epidemics should be frequent with the small spores pathogenic on broad leaves. This conclusion, too, is worth testing with a view to its eventual use in forecasting. Rain is a well-known cleanser of the atmosphere, washing spores out of the air and onto foliage and soil. This effectively increases p, the trapping efficiency. Rain has thus been an important factor in the consideration of long distance dissemination of spores: spores of, say, *Puccinia graminis* are carried aloft and then northward in a maritime tropical air mass, eventually being washed to earth by precipitation. The forecaster must bear in mind the dilution of the spore cloud with distance, but conceivably synoptic charts of the air currents at several levels could be useful in predicting the transfer of prolific fungi that infect large acreages.

V. Infection

The pathogen has avoided many pitfalls before alighting, but it now has another crucial step to accomplish: it must infect a plant. In our hypothesis the number of spores deposited per successful infection is K; infection increases as K decreases. The first prerequisite is that the propagules be alive. The forecaster will have investigated the hardiness of the pathogen and entered this into his "rules." For example, he will be confident that the hardy rust spores or apple scab ascospores or the virus safe within an aphid are alive, and that this step provides no opportunity for a forecast criterion. On the other hand, he knows that the spores of the downy mildews are susceptible to drought and that sunlight is a signal for a forecast of "no epidemic."

Germination is the second prerequisite for the success of a spore. This step is frequently susceptible to weather and has been a most fruitful source of forecasting methods. Satisfactory conditions for germination must arrive while the spore is viable. The requirement for moisture is common and has been widely applied in forecasting, e.g., the downy mildews. In contrast to this are the powdery mildews with thier requirement for high humidity but not water presumably, this could be employed in forecasting. The temperature is also critical during germination. Not only can the temperature itself be limiting, but it determines the speed of germination, and hence, the number of hours of satisfactory moisture conditions required. Thus, varying critical lengths of moist periods can be set for varying temperature.

Next, the landing and germination must occur on a host. This success is improbable if few of the host species grow near the source or if the plants are small and cover little of the soil. In this case few of the spores will alight on a host, and many will fall and few infect; the forecaster may be conservative in his warnings. Finally, the host that receives the pathogen must be susceptible. In the case of a heteroecious fungus, epidemics are most likely if the repeating stage is on the economic crop. Stakman (1942) has provided us with examples; *Gymnosporangium juniperi-virginianae* sporidia alone can infect apple, and no epidemic of rust would be forecast far from red cedars; *Puccinia graminis var. tritici* uredospores can infect wheat, and the pathogen can spread through a wheat belt by repeated, relatively short jumps with scattered sources appearing near many hosts.

In any case the strain of the pathogen must be capable of infecting the variety of the host. One of the forecaster's most important indicators is a survey of the various races present and the varieties being grown. Thus, the appearance of Race 56 of *Puccinia graminis var. tritici* and the larger acreages of Ceres wheat foretold the elimination in the midthirties of that hitherto resistant variety. The damage wrought by an infection could be the *D* of out hypothesis. If this were our thinking, then the proportionality constant *K* would be the number of propagules deposited per unit value destroyed. Then *K* would fall, and the probability of damage would rise if we were concerned with a systemic disease, particularly of a large host. The effect is increased further if the individual plant is valuable. Consider how the probability of damage rises as turn from the spots of apple scab to the systemic infection of Dutch elm disease or oak wilt of a prized shade tree. In the preceding sections we have dealt with the several steps in the multiplication of an air-or insect-borne pathogen.

The life of a soil-borne fungus may be less complicated, but an infection must be accomplished and the pathogen flourish within the host if an epidemic is to ensue. The separation of infection and incubation is incomplete in our knowledge but largely unimportant for this discussion. Seedling diseases are frequently associated with slow emergence and retarded early growth. Thus, the occurrence of damping-off of corn by *Pythmin* spp. can be forecast for cold, wet springs. The more specific, soil-borne wilt fungi also prosper best in certain environments: *Verticillium* wilt of tomatoes can be forecast for cool and *Fusarium* wilt for warm climates. Infection phenomena are often susceptible to the weather and provide a wealth of opportunities for forecast criteria. Can the pathogen survive, germinate, and infect in the environment of the host? If the forecaster answers these in the affirmative, he knows that a frequent barrier to epidemics has been removed and that he had best beware.

VI. The Incubation Period

The rapidity of increase of a pathogen has already been mentioned as crucial to the development of epidemics and consequential losses. This is so in large part because the pathogen and host rarely reach equilibrium during the life of an annual crop. Rather the pathogen is racing against the time when unfavourable weather will arrive. Therefore, what Chester (1950) has called "tempo" is critical. The length of the cycles as well as the successful infections per cycle determine the tempo.

The discharge, transfer, and trapping of inoculum and the infection of the host can, and often do, occur within a few hours. Incubation is generally a more extended process and considerable variation in its duration is frequent. Here, then, is an opportunity for alterations in the tempo, alterations critical to forecasting, alterations in the rapidity of reappearance of new sources and in their size Q. Within the fungus safely within the moist host tissues during incubation, temperature—not humidity—becomes the important factor. An example of the importance of temperature in the length of the incubation period is provided by *Phytophthora infestans*. Decreasing the maximum and minimum temperatures from 23 and 15^{0}C. to 20 and 10^0 C. increased the length of the period by one-fourth. This would result in only 5 instead of 6 multiplications of the pathogen per month. Of course, raising the maximum and minimum temperatures to 35 and 25°C. destroyed the pathogen within the host.

A similar example of altered length of period is provided by *Uncinula necator* : the incubation period was halved by raising the temperature from 15 to 26°C. *Peronospora tabacina* illustrates the disastrous effect upon the pathogen's prosperity of high temperature during incubation: hot weather causes "dry weather" blue mold lesions which are devoid of spores, and hence, ends the multiplication of the pathogen, Unquestionably many examples of accelerated multiplication by warm temperatures and truncated epidemics due to hot temperatures can be found among bacterial and viral as well as fungal incubations. If the forecaster is aware of these relations for his charges, he can estimate whether the tempo of multiplication is sufficiently rapid for an epidemic to develop within the season. A second use for a knowledge of the length of incubation is in detecting synchrony of the cycles of weather and pathogen.

Imagine that a period of weather unfavorable for sporulation and infection is followed by a day when infection does occur. If a second day of weather suitable for multiplication occurs before the end of the incubation, new infections will be add. If, instead, the second suitable day occurs at the end of the incubation, infections will be multiplied, no just added. Thus, favorable conditions alone are not enough: they must be meshed with the cycle of the pathogen, a cycle timed in large part by the temperature during incubation. One large class of pathogens need not race with the host to a deadline at the end of the season. These are the pathogens of perennial plants, such as trees. Here ample time is allowed for an equilibrium to be established, and eradication and exclusion are relatively harmless to the pathogen over the long pull.

The forecast will, therefore, be in terms of "when", not "if." The forecaster must guess the equilibrium rate from the best knowledge of the susceptibility of the host, the multiplication of both host and pathogen, and any deterrents to spread applied over the entire region concerned. As an example, among the American elms in the climate of Connecticut one can forecast with some confidence that in the neighborhood of 20% of the elms will become infected annually with Dutch elm disease if an insecticide is not applied, 3%, if it is. The incubation period is lengthy, relative to the other periods of the pathogen's multiplication, and hence provides a wealth of opportunities for the forecaster, who must estimate how the race between an annual crop and its pest will end or at what level the equilibrium between a perennial host and its pest will be reached.

VII. Integrations

The large number of possible prediction criteria enumerated above are not only opportunities for improving forecasts, but are also opportunities for hopeless confusion because of their number. How can such complexity be resolved? Fortunately a study of the biology of a particular pathogen will generally reveal a critical step to which attention can be directed. Chester's (1946) description of the forecasting of wheat leaf rust provides an excellent example. In Oklahoma, March is the critical month. If the weather during this period permits the first cycles or rust renewal, there is enough time during the ensuing season for the pathogen to reach epidemic proportions. After this month, the weather is rarely limiting. Therefore, the weather and rust behavior during this month can form the basis of a forecast of rust intensity 2 to 3 months later. The well-known Dutch and Irish rules for forecasting potato late blight also are a simplification based upon the existence of a critical stage in the pathogen's life. In the maritime climates of northeaster Europe the weather during incubation is rarely limiting, and adequate primary inoculum is apparently common. Therefore, the multiplication of the pathogen is limited only by the conditions for dispersal, transfer, and infection. Satisfactory conditions for these three steps are defined by simple rules, and hence, late blight forecasting is straightforward in Europe.

In the central United States, drought or hot weather frequently persists long enough to destroy *Phytophthora infestans* in the host. Therefore, Wallin (1958) deletes from his forecasts of blight the effects of earlier favorable periods if unfavorable weather persists for 21 days or more. The length of the incubation period can be the critical factor in epidemics. Shatsky's modification of Muller's incubation calendar for vine mildew illustrates this point. The incubation period is related quantitatively to the mean temperature, and fungicide spraying is indicated at the end of the periods thus estimated. An integration with a strong family resemblance to the above devices has been employed in Connecticut for several years for the prediction of tobacco blue mold epidemics.

The biology of the parasite has been reviewed by Stover and Koch (1951), and the spores are known to be air-borne in the morning.

Bearing this information in mind, the forecaster examines the hygrothemograph, sky cover, and rainfall records, together with a census of primary inoculum from the tobacco region and decides whether the pathogen could or could not have completed dispersal, transfer, and infection. If an infection period is thought to have occurred a "2" is entered on the calendar.

If two more periods occur before incubation of the first is complete, 2's are added each time and the numbers entered on the calendar are "4" and "6." If instead the second infection period occurs near the end of the incubation period of the first infection, the number is doubled and a "4" is entered on the calendar. When a drought or heat wave intervenes and sterile lesions appear, the number on the calendar is reduced to "1", and the multiplication process must begin anew. Thus, the forecasting scheme is an estimate of the population of *Peronospora tabacina* derived from a knowledge of its biology, its primary population in seedbeds, and the weather.

We have found in Connecticut that damage of consequence will appear in the field when the estimate on our calendar reaches "64." The arrival of this level can be anticipated from the tempo of the disease and the weather outlook. The preceding examples illustrate how the complexities of the pathogen's biology can be digested into rational forecasting criteria. Therefore, the prospective forecaster need not become lost in a maze on the one hand or rely upon empirical criteria upon the other hand.

VIII. The Usefulness of Forecasting

When one speculates about the value of weather, yield, or disease forecasts, no difficulty is encountered in convincing oneself that forecasts are useful, even necessary. The desired to foretell the future is strong. Nevertheless an enumeration of the characteristics that determine the practical usefulness of predictions is healthy, especially if it is made before a forecasting system is initiated. The first requirement of a forecast is that it be correct. Perfection is rarely possible in forecasting, and the natural tendency is to be conservation, to "overforecast." In disease forecasting this leads to the warning "epidemic" whenever a chance of such exists. At the same time that the forecaster is being "conservative", the growers' other advisors are naturally enough behaving in the same manner: they urge the grower to exercise control measures at all times. Therefore, the usefulness of the forecaster's warning of "epidemic" may be limited because his is but one more voice added to the chorus of voices urging him to repent and spray.

On the opposite side, the forecaster may be more helpful. He may lead the grower to omit costly control measures by forecasting "no epidemic." Before he can persuade the grower to make such a savings he must have established the accuracy of his predictions in the minds of the grower and his advisors and have overcome the natural tendency to be conservative. In France this happy state has apparently arrived: the warning service for vine mildew has led to the omission of two sprays in most years at an annual savings of two and one-half billion francs. Thus, if the forecaster can bring about the elimination of unnecessary and costly measures through accurate forecasting, he can be truly helpful.

The distance that the forecaster can see into the future, as well as his accuracy, affects the usefulness of his methods. If the prediction arrives in time for remedial action to be taken, it is helpful. If it arrives too late for action, it only lengthens the time over which the grower must bear his sorrow. The predictions of epidemics of seed-borne smuts and virus from the conditions during the flowering of the wheat or form an index of potato tubers has already been mentioned. Here the prognostication can be made for many months in advance, and the grower has ample time to remedy the matter.

The predictions of wheat leaf rust and of bacterial wilt of sweet corn extend for several months into the future. Hence, the farmer has time in which to benefit from the forecast by planting an alternate crop or variety. The French warnings concerning vine mildew are of much shorter period; nevertheless, they arrive in sufficient time to be helpful because fungicidal control can be applied or omitted on short notice.

Long-range or climatological predictions plant diseases have been made informally. Men have recognized that certain climates are conducive to a disease, others are not. Seed production has been located in regions where the climate makes important seed-borne disease unlikely. In addition to this a formal study of long-range or climatological prediction would be interesting. The epidemics of certain diseases can be traced to critical weather events. The probability of these events can frequently be estimated from the series of weather observations that have now reached great lengths in many localities. The knowledge of biology and weather could be combined to produce estimates of the probabilities of disease at various localities and seasons. These "forecasts" should prove useful because of their long, almost infinite, extension into the future. The discussion of the period of the forecast has brought out the importance of the control method. If the forecast is to provide more than intellectual satisfaction, there must be a control method that can be exercised and the impending epidemic forestalled.

The usefulness of the above prediction of smut and virus epidemics depends upon the existence of alternate, clean seed. The usefulness of the predictions of wheat leaf rust and bacterial wilt of corn depends upon the existence of profitable alternate crops or resistant varieties. The benefit from the warning systems for downy mildews is dependent upon the efficacy of fungicides. Two diseases stand in contrast to the ones we have just examined. Apple scab infection can be eradicated by suitable chemicals. Hence, a census of infection can be used and the need for a forecast is consequently less. Wheat stem rust control at the present depends practically upon the choice of a resistant variety. Consequently, the forecast of a stem rust epidemic issued in the middle of the growing season is nearly useless to the individual farmer because there is as yet no remedy he can apply to forestall the impending calamity. The use of a forecast necessitates changes in operations. Flexibility is a prerequisite to the employment of prediction. At least two things affect this flexibility.

The first is the quantity of labor and equipment available for control measures relative to the acreage that must be treated. If the entire area can be treated in a day or two, treatment can be delayed safely until a forecast is issued. If the entire area requires many days for treatment, however, the machines must be kept in operation continuously; otherwise, a large proportion of the area might be damaged before treatment in response to a warning could reach it. The second thing that affects flexibility is the complexity of the farming organization. If the man who applies the treatment is also the man who decides when to treat and if supplies are readily available, a treatment can quickly follow warning. Alternatively, if a large organization is involved, planning must precede action. Here the confusion that would follow a change in plans following a warning might well cost more than the application of an occasional unnecessary treatment. Consequently, a short-range forecast is not highly useful; perhaps climatological "forecasts" would prove to be.

An important—perhaps the most critical—factor in the usefulness and in the adoption of forecasting is the ratio of benefit to cost. How great is the benefit from disease control relative to the cost of the control measure? If this ratio is large, the rational grower will apply controls whether the disease is forecast or not. If this ratio is small, no amount or urging will induce the rational grower to act. In the northeastern United States apple and potato growers know that the yield, appearance, and keeping quality of their produce are profoundly affected by apple scab and potato late blight. They also know that including an effective fungicide among the pest sprays they are already applying is inexpensive. Thus, the ratio of benefit to cost is high, and the growers tend to apply fungicides routinely with little regard for the likelihood of an epidemic.

At the opposite pole stands the owner of a large wood lot populated in part by elms. The death of the elms would not be disastrous to the production of the lot. The cost of an annual application of an insecticide to kill the vectors of the pathogen of Dutch elm disease would be large relative to the annual production of the lot. Therefore, the ratio of benefit to cost is small, and the owner is unlikely to apply a control measure, even when he is assured of the imminence of an epidemic of Dutch elm disease. Between these two extremes lies a region where the benefit: cost ratio is favorable to the employment of forecasts. The example of the vine mildew warning system in France cited above lies in this region of a favorable ratio. The losses from mildew are disastrous. On the other hand, considerable savings can be made by eliminating unnecessary sprays. Hence, the warning system is employed by the growers.

The benefit: cost ratio also depends upon the annual variation in the amount of disease. If the amount is nearly constant from year to year, experience has taught the farmer how to make his own forecast: next year will be like this year and will have the same benefit: cost ratio and demand the same measures. This leaves little for the forecaster to do that is useful. When the amount of disease is variable from year to year, the forecaster can be useful, for he will predict when the benefit of treatment will be great or small relative to the cost of treatment. The forecasting of disease is a contribution to the prediction of yields. It logically proceeds from a knowledge of the interaction of host, pathogen, and environment. In this review the etiology of disease has been examined for opportunities for prognostication of disease and subsequent loss.

The magnitude of the primary inoculum affects the subsequent epidemic. As a clue to the size of the epidemic it has the advantage of earliness, the disadvantage being subject to many influences before harvest time. The production of inoculum and its dispersal into the medium about the plant is one of the processes upon which weather can exert its influence and is, therefore, a rich source of forecast criteria. The transfer of inoculum by air and soil, and as a contaminant is for the pathogen a risky business, because it is dependent upon fortune. The forecaster profits from a knowledge of the proximity of the source of inoculum and of the number of vectors during this random process. Propagules are trapped and their numbers depleted more rapidly by some types of foliage and weather than by others. This is probably important in predicting the distribution of infection about a source of inoculum.

Infection is a late step in the pathogen's multiplication, but its requirements are sufficiently critical to provide a wealth of criteria for prediction. The general requirement for liquid water at this stage provides the basis for several classic forecasting schemes. The incubation period is generally much longer than the preceding steps, and therefore is subject to changes of length. Thus, it can influence the rapidity of development of the disease and the level attained before unfavorable weather or harvest ends the annual race between host and pathogen. The myriad influences can frequently be digested into simple forecasting rules because of the limitation placed upon epidemics by a few critical stages. The usefulness of the prediction of epidemics depends upon the accuracy and range of the forecast, upon the existence of remedies and the ease with which they can be applied, and upon the benefits from disease control relative to the cost of remedies.

7

PREVENTION OF DISEASES

Here we shall discuss the physical and chemical therapy of the diseased plant; the treatment of the diseased plant as an individual, not the plant as a member of a population. We shall consider the plant as a "patient" not as a public health charge. Protection is aimed at the healthy plant. Therapy is aimed at the diseased plant. Protection is aimed at the pathogen as it lives and moves between hosts. Therapy is aimed at the pathogen after it has arrived in and has established housekeeping in the host. Therapy is the cure of a sick plant, the mitigation of its symptoms, or the repair of the damage whether the pathogen be animate or inanimate. The practicing plant doctor is interminably being presented with sick plants, singly or *en masse*. He usually gives the disease a name on the basis of symptoms, identifies the fungus on the basis of its structures present, and suggests a preventive measure.

The grower too often still has a diseased plant and a confused mind, with the one clear, happy thought that he paid no fee—at least directly. In the literature written for the advice of growers 50 years ago, "cure" of plant diseases meant to "get ahead of the enemy." The reader was told that internal fungi, when once they are established, must be treated by removal and burning of diseased parts. This is common advice today, though mortal for the suffering plant. As our knowledge of the biochemical nature and mechanisms of disease increases, there is more concern that the plant patient lives. The next decade or two should see a "breakthrough" in the therapy of the diseased plant. This phase of plant pathology is still in the adolescent stage. The framework of outlined, but the ideas and applications that are necessary to reach full maturity are appearing here and there in the writings of many men. An attempt has been made to encompass and clarify today's concepts on plant therapy with the hope that progress can advance faster in this important field. A plant is today thought of as a dynamic entity maintained in health by a correlated series of interdependent metabolic processes. Animate or inanimate factors may disturb the usual functional pattern and result in developmental changes.

Therapy removes the cause of disease so that the normal mechanism of living can run smoothly. Combating the causal agent after it has injured the plant is a form of disease control labeled by some as "disinfection." However, the concept should include not only the neutralizing or stopping of the injurious stimulus, but also the repair of the cellular damage done. Hence, the subject matter concerns the pathogen-host interactions during the time interval between "infection" and death of the plant or its organs—a lifesaving action. Knowledge of the pathogen-host relationship must answer many questions before effective therapy can be carried out. Usually the extensiveness of the infection is first thought of in order to make use of physical means of therapy aimed at confining the zone of injury.

PHYSICAL THERAPY

Two courses are open for the therapy of plant diseases, physical therapy and chemotherapy. In the former, the disease is fought with physical means, such as surgery, temperature modification, radiation, and moisture modification. In the latter, the disease is fought with chemicals that act typically or systemically.

Surgery

Surgery is the removal of infected tissues to prevent additional damage. Sometimes the plants themselves are able to stop further injury by shedding diseased structures. The "shot hole" symptom in leaves of cherry and poplar is familiar to pathologists. Similarly, leaf cast may avoid infection of the twigs, petal fall may avoid involvement of the fruit, and fruit drop the involvement of the spur or stem. Self-therapy is accomplished by the shedding of infected bark where the growth of submerized tissues exceeds the rate of penetration of the pathogen through the cortex. While expressed physically, the underlying mechanism is biochemical. An unusual therapeutic measure has been proposed for the American leaf spot ("Ojo de gallo") of coffee by Wellman (1950). Gemmae of *Mycena (Omphalia) citricolor* serve as the inoculum and are too heavy to be borne by air currents, requiring dissemination by splashing rain. Hence, complete defoliation of the diseased trees rids them of the fungus.

A chemical "pruning" may be effective for ridding tomato plants of the root knot nematode. While evaluating the nematocidal activity of chemicals, Hopper and Tarjan (1954) observed that following chemical treatment of the soil with *p*-chlorophenyl rhodanine around young infected tomato plants they grew normally. Careful examination revealed that the infected roots were killed, but the plants formed a new root system free of nematodes. Man uses, in many forms, the principle of mechanical excision of diseased parts as a means of therapy. Removal of localized knots on plums and cherries, and galls on olives and roses are among the more common examples.

The systemic wilts are sometimes checked by pruning infected roots or infected branches. Cutting off branches of pear trees several inches below the zone of visible fire blight symptoms is practiced to halt further invasion by the causal bacteria. We have found that removal of the fruits and succulent cane tips of rose bushes stops the advance of *Botrytis* downward through the pith into the crown. The bark over diseased areas in tree trunks and branches is removed to permit a drier environment unfavorable to the pathogen, e.g., slime flux. For a similar purpose, bark is excised from diseased areas in the exposed main roots and crown of walnuts and chestnuts. Sometimes it is done to permit better contact of toxicant with the pathogen. Bark tracing or removal has been used to check bleeding canker of hardwoods and stripe canker of *Cinchona*. However, the causal *Phytophthora* spp. remain in the roots and may grow upward to kill new sections of the cambium and adjacent living cells.

Temperature Modification

A well-known measure for freeing plants of pests is based upon the differential heat inactivation of the pathogen in, or on, the host. Generally the extent of the injury depends upon the physical state of the respective protoplasm, the degree of hydration or solation. The composition of the protoplasm and the containing membranes also affects the action. When possible, the life activities of the microorganism to be eliminated are stimulated and the host is maintained dehydrated and/or dormant. Occasionally, the curing of diseased plants by growing at above-normal temperatures is reported. Brierley and Smith (1957) observed that White Wonder and Dynamo chrysanthemum plants infected with the flower distortion virus, when grown in a greenhouse maintained at 35^{0}C for 2 to 3 months, produced tip scions free of the virus. Hot water immersion, with or without antiseptics in solution, has been used to free crucifer seed of viable black rot bacteria, strawberry plants of virus, and bulbs and rhizomes of nematodes. The differential in heat resistance of host and pathogen may be explained on the basis of the heat stability of their respective enzymes. The thermal growth range or organisms reflects the thermal stability of their cellular proteins. These in turn are affected by the cell water content and the physical and chemical state of the cell components.

Radiation

Ultraviolet radiation is often proposed as a means of plant disease therapy. The researches of Fulton and Coblentz (1929) sum up its advantages and limitations. The shorter wavelengths (about 240 mμ) exert the greatest germicidal action, which decreases with wavelength until about 365 mμ which is an upper limit. The abiotic effect varies with the time of exposure and the intensity of the light. The principal limitation seems to be the inability of the ultraviolet rays to penetrate sufficiently beneath the surface to destroy the thalli of pathogens. Infrared energy has been exploited for killing microbes in foodstuffs and materiel, but seemingly is not sufficiently selective to be used in plant therapy. Instances where exposure of plants to the visible light spectrum (313 mμ and longer) has suppressed disease, leave in question whether a direct lethal effect, or the indirect change, is responsible. Ultrasonic inactivation of microorganisms in and on inanimate substrates has been shown, but reduction to practice on living plants, is as yet, not practical.

The therapeutic value of radioactivity as well as its harmful effects on animals are well known. Solutions and powders containing alpha particles were distributed for test on plant disease by one manufacturer, but no beneficial therapy was evident against Dutch elm disease in our tests. Beta and gamma radiation exert potent stimulatory and lethal action on cell protoplasts but, as yet nonselectivity, cost, and danger to the user militate against their use in plant therapy. Scientists of the U.S. Department of Agriculture, according to a recent news release, had hoped that nursery stock and other agricultural products could be freed of plant-pathogenic nematodes by exposing them to ionizing radiation. The golden nematode of potato (*Heterodera rostochiensis*) can withstand radiation up to 20,000 roentgens before the females are sterilized and 120,000 roentgens or more are required for complete killing.

Man is a "softie" by comparison, since a mere 300 to 650 roentgens are fatal. Other nematodes require between 350,000 and

640,000 roentgens for a lethal dose. So therapy e is little hope that radiation can be used for killing nematodes in living plants, because nematode-killing doses of radiation also injure plants. The possibility of using ionizing radiation as a therapeutic agent for a bacterial disease has been explored. Crown gall, caused by *Agrobacterium tumefaciens*, is suppressed by irradiating the inoculated plant with X-rays. Irradiation usually does not kill the pathogen but rather destroys the auxin system of the plant, so that it is unable to respond by gall formation. Thus maleic hydrazide, which also prevents growth of the plant, inhibits gall formation in inoculated tomato plants.

Generally, the susceptibility of a fungus pathogen to ionizing radiation is much lower than that of the host. Waggoner and Dimond (1957) have studied these interrelations together with the changes in resistance to disease undergone by tomato plants on exposure to X and gamma radiation when plants are then inoculate with *Fusarium oxysporum f. lycopersici*. The plants were killed with lower doses than the fungus. Generally speaking, the effect of radiation treatment on susceptibility of the plant is related to the effect of irradiation on the auxin system. Potato tubers, irradiated to delay sprouting are more subject than control tubers to rotting by *Erwinia atroseptica* and *E. carotovora* because the treatment delays or prevents periderm formation.

Moisture Modification

Intercellular humidity is a factor concerned with disease in a large number of plants. This is most likely where the pathogens are primarily intracellular invaders. Hence, the extent of spread of such diseases within a plant, may be limited, when controllable, by the amount of moisture made available in the environment. This effect of intercellular humidity was demonstrated by Shaw (1935) to exert a definite influence on the degree of fire blight susceptibility in plants of pear and apple. Data indicate, that when 99.5—100% intercellular humidity occurs, maximum host damage results. Conversely, when a turgor deficit, or intercellular humidities of 96% or less obtain, no disease develops. This idea has been validated by Clayton (1937) with another bacterial disease, blackfire of tobacco. The modification of the moisture-oxygen content of woody stems acts in the therapy of *Verticillium*-infected maple trees by mitigating the symptoms.

Caroselli's data indicate that the degree of injury caused by the fungus is dependent upon the relative amount of moisture and air within the tissues of the tree. When the tree is in full foliage, the water content is least and the air content greatest. Presumably, the additional oxygen favors growth of the systemic fungus so that the wilt symptoms are exhibited after the next rain. A modern, large-scale commercial application of this measure is the "vacuum cooling" of lettuce and other food crops.

The plants are exposed to a partial vacuum. This converts the interior liquid water films to intercellular water vapor which is then withdrawn from the plant structure. This vaporization cools and dries the intercellular spaces. Although thought of as prophylaxis by cooling, a sufficient intercellular drying results to retard the development of bacterial soft rot and other diseases. Thus, Shaw's (1935) hypothesis is reduced to commercial practice.

Intercellular moisture may be manipulated in plants to minimize the extent of disease. Powdery mildew diseases are prevalent in sites where water deficits due to transpiration exceeding absorption often occur daily or for short periods. A possible explanation is that the haustoria of the extramatrical mycelium are unable to invaginate protoplasts exerting a positive pressure. Thus, by consistently making adequate water available and reducing water loss by one of several means, the spread on mildew can be checked. A further example of moisture modification and attendant disease damage is the case of golf-green turf and *Curvularia* blight. This may be as much prevention as therapy.

On hot, mid-summer afternoons, the close cut grass (1/4-inch) wilts due to a water deficit in the leaf blades. This predisposes them to attack by the fungus, which is a common black mold growing on turf debris. Maintaining the turgidity and vigor of the leaf cells by applying a light watering in early afternoon appears to check the advance of the facultative parasite into the healthy tissues. Endoxerosis or internal decline of lemon fruit was found by Bartholomew (1928) to be caused by exposure of the trees to high temperatures with consequent rapid leaf transpiration when relative humidity of the surrounding air is low. Water is lost from the leaves and in turn withdrawn from the fruit to the point where cellular damage and gum formation occurs. Endoxerosis is most prevalent when growth activities are greatest.

Hence, therapy is aimed at maintaining an intermediate rate of growth by manipulation of contributing environmental factors. Blossom-end rot of tomatoes is a similar disease. Likewise, any practice which tends to conserve moisture and provide a uniform supply to the foliage and fruit reduces losses. Aging celery seed for 3 years often results in "die-out" of *Septoria*. This may well be due to the drying of the seed.

TOPICAL CHEMOTHERAPY

Chemotherapy is one of the exciting new frontiers of plant pathology. Chemotherapy may be topical or systemic. Ever since Prevost (1807) treated wheat seed with copper sulfate, research on chemical protection of plants against invasion has been going on. Success has been nothing short of fantastic. We have had Bordeaux mixture, wettable sulfur, "fixed" copper fungicides, choranil, ferbam, nabam, zineb, dichlone, glyodin, captan. The list lengthens daily. These have been written, reputations made. Similarly, we have pursued the microbes in the soil with formaldehyde, carbon disulfide, chloropicrin, ethylene dibromide, and 1, 2=dibromo-3-chloropropane. This list is shorter, but it lengthens, too. Thus, we have protected our crops. Thus, we have killed the microbial pathogens during the inoculation stage. During all this time, we have neglected internal therapy. Chemotherapy now gives us hope in the attack on diseases where chemical protection has failed or is impractical. Most of the modern research is aimed at vascular diseases, at virus diseases, or at bacterial diseases. Few of these have receded before the onslaught of protectants. Perhaps, they will yield to chemotherapy!

Several general and specific discussions on the chemotherapy of plant disease have appeared during the past decade. The potentialities have been covered by Stoddard and Dimond (1949), Crowdy (1952), and Brain (1952). A compound that initiates a curative or mitigating effect, either directly or indirectly, is termed a "chemotherapeutant." Perhaps an ion or an atom may as well be the active agent as a compound. The pathogen may injure only surface cells or a localized area of tissue. On the other hand, it may systemically injure the entire plant or any major part thereof. Treatment of the former is topical chemotherapy. Treatment of the latter is systemic chemotherapy. Therapy will advance as we learn how "toxins" act, whether by destroying tissues or by inhibiting or stimulating essential metabolic functions.

A key point is the initial reaction that takes place between the invading organism and the host tissue. Such knowledge would permit defining more accurately the specific cellular functions disturbed during the infection period, and would perhaps suggest a therapeutic measure; how to selectively neutralize the pathogen or its metabolic products within the tissues of the host. Chemotherapy has contributed its part to the subtle realization that only minute concentrations of some chemicals are required to kill fungi and stop disease in plants.

Painstaking laboratory techniques and the use of the log-probability curve as a device for estimating the microbial ad plant toxicity of compounds have also contributed their part in changing our ideas of biological activity from pounds per hundred gallons of water to parts per million.

Therapeutic Index

Ehrlich of salvarsan fame was surely the founder of chemotherapy as a direct killing action. He emphasized very strongly, indeed, the principle of the chemotherapeutic index; namely, that the dose of the drug needed to kill the pathogen must be below that to kill the patient. The wider the ratio, the bigger the index, and the safer the treatment. In the plant pathologists' jargon, the chemical must not be seriously phytotoxic. This principle dictates the screening techniques to be used and phytotoxicity tests must come early in the procedure. The literature on the inner therapy of plants has been summarized by Muller (1926). In addition to his own experimentation, there is assembled an orderly, comprehensive account of other investigations on internal plant therapy before 1926. He repeats the idea that "therapy" is direct control, particularly useful for living endopathogenic and inanimate causal agents. He evaluates chemotherapeutants according to a "Therapeutic Index", which represents the curative dose divided by the tolerated dose ($I = c/t$).

Factors affecting the curative dose are pointed out as: (a) kind and condition of the therapeutant, (b) duration of absorption period, (c) desired degree of saturation, and (d) local conditions, i.e., time of year, temperature, air movement, relative humidity, and cloudiness. Factors affecting the tolerated dose by the pathogen (i.e., kind, state of development) and by the plant (i.e., family, morphology, volume stage of development) are thought of in relation to the strength of the outbreak, the kind of infestation, and the site of the disease. The topical application of medicaments to surface lesions in order to reduce infection has been practiced for centuries; even though they have been only dung, urea, lime, sulfur, or charcoal.

The caustic action of lime sulfur was found to "burn out" the scab fungus (*Venturia inaequalis*) on apple leaves, but often it injured the plant tissue also. Perhaps the largest commercial success of such therapy is the cure of apple scab with water-soluble organic mercurials, introduced as phenyl mercury triethanol ammonium lactate ("Puratize") by Howard and Sorrell (1943). Another commercial success is the curing of cereal seed invaded by the smut fungus. Success has been reported for the cure of cedar rust by Strong and Cation (1940) with sodium dinitrocresylate. Ark (1941) successfully treated bacterial crown gall with the same chemical. Therapy by the surface treatment of tree branches with chemical that permeate the bark ahs been practiced for many years. A zinc chloride-glycerine-alcohol preparation has been used in California for the control bacterial blight of pear. Coal tar oils are applied in Poland to check brown rot (*Sclerotinia* spp.) cankers on stone fruit trees.

Vegetative parts of plants used for propagation are commonly dipped or soaked in solutions of antimicrobial agents to rid them of established pathogen s. An example once widely used is the treatment of seed potato tubers by immersion in solutions of formalin or mercuric chloride. While fairly effective as a curative measure, the phytotoxicity is high and the general presence of *Rhizoctonia* sp. and/or *Streptomyces scabies* in fertile soils militates against its use. The seeds of many plants are infected internally with bacteria or fungi. Plain water alone is credited with curing disease. Tyner (1957) reported that loose smut, caused by *Ustilago nuda*, can be eliminated from barley by water soaking. He suggests that the effect is not due to microorganisms or their chemical products accumulating in the soak water, but rather to quinones formed in the germ of wheat or barley during the soaking. Accelerating the healing of wounds by stimulating callus formation through surface application of chemicals has been attempted for many years. Lanolin was found to improve the initial stimulation of healing by Shear (1936) and was further enhanced by addition of such growth-regulating compounds as 4-chloro-3, 5-dimethylphenoxyacetic acid and 2, 4-dichlorophenoxyacetic acid. Crowdy found little evidence that the duration of the healing period can be extended markedly by chemical treatment or that out-of-season healing can be stimulated in this way. The improved healing may be due either to an earlier start of the healing process, or to an accelerated rate. Timing or chemical treatment with normal growth activity of the plant seems very important.

Systemic Chemotherapy

General Concept

Internal systemic medication is the real frontier of therapy. This challenges us so much because it offers hope of chasing a pathogen into the farthest leaflet of the plant and killing it. The entomologists are ahead of us here. They can kill a leaf miner in the highest leaf of a birch tree by spreading an ounce or so of an organic thiophosphate over the soil in the spring. We must some day be able to repeat this astonishing performance for the Dutch elm disease. Perhaps we can console ourselves in our slowness by saying that the thiophosphate acts as a nerve poison on the insect larva and that the trees have no nerves. Our fungus, *Graphium ulmi*, has no nerves either and, hence, is not damaged. We must not forget the ancient and hoary principle in human medicine that every disease has its specific remedy. We researchers in chemotherapy are aiming at that objective. This subject has been neglected hitherto; because individual plants have been considered expendable, because the problem is intensely complex, and because we did not have access to the formidable array of chemicals and antibiotics now available to us. In this chapter we shall not be describing, however, very many practically useful chemotherapeutants. We shall concern ourselves with the nature of the problem, with some of its pitfalls that we have noted, and with some of the challenges. Other facets of the subject have been considered recently by Dimond (1959) and by Dimond and Horsfall (1959). We hope to suggest that "the water is fine" and to encourage others to try a swim. We hope to shed some light on the problem posed by the old hymn: "Watchman, tell us of the night, what its signs of promise are."

Nutritional Imbalance

Between therapy by physical measures and what is generally thought of as chemotherapy, is the well-established principle of correcting nutritional disorders of the plant cell. Most commonly, plant disease is thought of as being caused by animate pathogens. However, therapy can and has been, consciously or unconsciously, applied to correction of inanimate causes of disease—those nonliving factors that affect adversely the natural or normal functioning of plant cells, namely, imbalance of water, oxygen, and nutrient elements. Balancing certain elements in the soil environment or in the plant cells can be used as a therapeutic measure. Tomato fruits low in calcium but high in total nitrogen, iron, and copper are more liable to blossom-end rot according to Taylor and Smith (1957). Conversely, supplying the plants with more calcium and decreasing nitrogen levels has reduced the disease.

Similarly, the use of sulfate fertilizers rather than chlorides can reduce succulence of tomatoes and hence mitigate damage. This is therapy of nutritional imbalance. The major nutrient elements necessary for metabolic equilibrium, viz., nitrogen, phosphorus, potassium, calcium, and magnesium, are taken care of by adjustment of fertilizer practices. The minor elements that may induce the so-called "deficiency diseases"—zinc, boron, copper, manganese, sulfur, and iron—are identifiable from a diagnostic key to symptoms prepared by McMurtrey (1948). One has the option of considering as chemotherapeutants the metabolites or nutrients named above, when used to maintain or to improve the normal cellular activities. After observing the therapeutic effect of zinc on mottle-leaf of *Citrus*, Reed and Dufrenoy (1935) explained the action on the basis that the salts of certain metals catalyze the partial oxidation of sulfhydryl compounds. Certain of these compounds are present in all living cells and control the life through maintaining the energy at a given level by oxidation. This level is defined as the oxidation-reduction potential; which when low, may result in pathological symptoms correlated with an accumulation of suboxidized metabolic substances as in mottle-leaf of *Citrus*.

Profound changes in the cytological conditions are associated with the recovery of mottled trees after then entry of zinc, applied either to the soil or to the foliage in the form of a spray. Zinc accumulates in the meristematic cells of buds and in the palisade cells of green leaves. After zinc sulfate has been applied, the symptomatic trees resume normal leaf cellular activity as judged by normal nuclei, fibrillar cytoplasm, and normal chloroplasts, and as exhibited by accelerated growth of new shoots. These effects suggest that some reaction has been initiated by which the proteins and carbohydrates of the cells have been utilized to supply energy to the cells. Steroids accumulate in cells of mottled citrus leaves, but are scarce in leaves to which zinc is applied. Seemingly, the stabilizing of the sulfhydryl compounds by zinc promotes the oxidation of cell metabolites and thereby liberates energy for vital processes.

Undoubtedly, zinc is linked with sulfur compounds in maintaining normal green plant cellular functions. Yet, in solution, zinc at concentrations of 5 to 25 p.p.m. has been reported to stop the growth and kill corn and citrus seedlings. How the zinc requirements of plants can be met without toxicity has been established by Mase (1914). Presumably, in the presence of calcium carbonate, zinc is precipitated as an insoluble salt. Necessary quantities are dissolved out by root excretions and absorbed as required by the plant without toxic accumulation. Thus, zinc carbonate or a similar basic salt would serve as a "safe" or sublethal reservoir of the essential toxic metabolite. This same principle can and has been used with other metal ions in other "deficiency" diseases of plants.

Molybdenum in very small amounts has been reported as needed for the physiological reduction of nitrates in plants. Nitrates accumulate in the tissues of molybdenum-deficient plants, and the protein content of the plant is reduced. This upset metabolism results chiefly in characteristic hypoplastic symptoms, as for example "whiptail" of crucifers. Treatment with a few parts per million available molybdenum suffices to cure this growth condition. In contrast, an excess of a single nutrient or a combination of nutrients may directly or indirectly result in injury. Here we get into the area of chemical antagonism and physical changes in the state of the plant protoplasm. Of major importance is the accumulation of sodium salts and resultant "alkali" or "salt" injury. Nevertheless, the removal of an excess injurious salt or element, even modification of soil reaction to within the range for normal plant growth, is a chemotherapeutic practice in the broadest sense.

Healthy living cells are believed to have an approximate balance of hydrogen and hydroxyl ions. They are able to maintain an internal neutrality because of their slight permeability to these ions. The internal pH can be upset by the greater penetration of undissociated molecules of acidic and basic substances than of their corresponding ions. Hence, when the plant cells are exposed to low pH values, weak nonionized acids enter the cells and damage them. Similarly, weak poorly dissociated bases are toxic at pH values above 7.0 because they can permeate the cell at that pH. Hence, correcting the chemical environment will cure the harmful effects.

Feldman et. al. (1950) observed that the Dutch elm disease organism (*Graphium ulmi*) forms its most toxid metabolite when the pH is 4.2 and successively less as the pH of the medium is raised to 7.0 or above. Therefore, an alkaline formulation containing lime was developed for the impregnation of soil around elm trees, to suppress or avoid the development of symptoms. A check *in vivo* of the pH of the tree sap indicated a rapid rise in alkalinity for a few days following soil application and then a dropping back to normal. The experimental results suggest that some curative effect was attained; either from the pH modification, or the high calcium hydrated lime and other ingredients used.

Biochemical Specificity

The idea of finding a chemical that may be used as a systemic medicine is based on the principle of biochemical specificity as instanced above for the birch leaf miner. The idea of biochemical specificity, perhaps, goes back a century to the recognition by Louis Pasteur that microorganisms are so specific in their biochemical reactions that they can separate optical isomers. Ehrlich (1913), 50 years later, seems to have had in mid selectively active substances when he established the term "chemotherapeutic." Dubos (1958) has stated recently that "under natural conditions, each one of the microbial species concerned in the economy or organic matter, is more or less specifically adapted to the performance of a limited, defined biochemical task." The concept of chemotherapeutic specificity has been sharpened over the years. Formerly it was considered on the basis of the whole cell, later on the basis of the enzymes, and more recently on the basis of the molecular structure of the substrate.

Dubos (1945) has pointed out that effective therapeutants do not behave as gross protoplasmic poisons which affect indiscriminately the structure and function of all living cells. They interfere selectively with some specific steps concerned in the nutrition, synthesis, or cell division of the pathogen. Other therapeutic agents may react selectively with well-defined structural components essential to the pathogenic behavior of the cell. This attitude is hopeful for the inhibition of pathogens within the tissues of the host plant.

Yarwood (1955) has presented evidence of "selective accumulation." Although the lethal dose of fungicide per unit of tissue be the same for host and fungus, the fungus may accumulate the toxic agent in larger amounts than the host, and is, therefore, killed at a lower applied dosage than the host. Autoradiographs indicate that this is true for sulfur therapy of rusts. Also there may be "selective toxicity." The amount of fungicide accumulated per unit of tissue may be the same for host and fungus, but the fungus is killed at much lower accumulated dosages than the host. This seems to apply to the sulfur therapy of powdery mildews. The possibility of finding chemotherapeutants exerting a specific action is strengthened by what has been learned from penicillin. Penicillin presumably blocks the synthesis of a particular wall component of the bacterial cell, but not that of the host. Thus, its action is specific. If growing bacterial thalli are unable to form cell membranes, they undergo osmotic lysis.

Undoubtedly, compounds with similar action are waiting to be found for plant chemotherapeutants. Enzymes capable of specific lysis of fungous structures are sometimes to be found among fungi. As yet, knowledge of these enzymes has not been developed for use in plant therapy. The plasmodia of some Myxomycetes have been observed to destroy the hyphae of fungi with the same apparent speed that a hot iron may cause nylon or acetate fibers to melt. The inky cap mushrooms demonstrate the property for all to see. Perhaps a better suggestion of possibilities for future specific agents is offered by the diplont or dicaryon stage of certain Ascomycetes. In the so-called sexual phase, a changed metabolic pattern follows differentiation of "fertile hyphae", an ascogonium, and ascogenous hyphae from the somatic tissues of the thallus. This diplont exerts a cannibalistic action on the haplont matrix of the surrounding stroma. This action has been used as an ordinal taxonomic character. It seems possible that the lytic enzymes, produces by the diplonat to make room for itself in the body of the haplont, could be isolated and provide the key information required to discover a specific chemotherapeutant useful in plant therapy.

Microbial Disease Therapeutants

1. *Intake or Entry*. Clearly we must start with the problem of intake or entry of chemotherapeutants . Our medical colleagues call this "route of administration." Through what routes can systemic therapeutants enter the plant? There are really only three practicable routes of administration for plants—root, stem, foliage. Mechanical injection is possible through each. Roach (1939) has published a basic treatise on the injection of materials and their subsequent distribution in the plant. Each portal of entry has its advantages, each its drawbacks.

Roots grow in soil. A compound aimed at roots must go first through the soil and be subjected to its severe substractive influences: distance, adsorption, direct chemical alteration, and biological degradation. The significance of distance is obvious. The farther the compound must go before it can reach the roots, the less will reach the roots. This follows from the simple laws of diffusion and dilution and is not of necessity related to any loss along the way due to hazards of the route.

Adsorption is a much more serious source of trouble. If a substance bears a strong charge, it will be adsorbed onto soil particles, presumably by van der Waals forces, and it may never arrive alongside the root to take it. Streptomycin bears such a charge according to Siminoff and Gottlieb (1951) and, thus, it is not an effective chemotherapeutant when applied to the soil. The same is true of thiolutin according to Gopal Krishnan and Jump (1952) and of numerous other antibiotics according to Martin and Gottlieb (1955). Ark and Alcorn (1956) showed some specificity in adsorption. Streptomycin is adsorbed more tightly be bentonite than by pyrophyllite. When Ark and Alcorn changed the charge on the clays with K_2HOP_4 the streptomycin was released.

Since soil is a reasonably reactive medium, it decomposes many candidate chemotherapeutants and reduces them to impotence. Our knowledge of this is pretty limited, but what we have is clear. Antibiotics seem to have provided more information to date than other types of therapeutants. According to Jeffsreys (1952), many antibiotics such as albidin, gliotoxin, and viridin come apart in soils where the pH is unfavorable. Even if the pH is not unfavorable, some antibiotics such as penicillin and streptomycin break down.

The soil is so full of microbes that it is not surprising that they, too, degrade chemotherapeutants very rapidly. Very few molecules have demonstrated radical resistance to microbial decomposition. DDT is an example, but, of course, it is not antimicrobial. Very probably the two are related. If DDT reacts with microbes to damage them, then the microbes will react with DDT and damage it. Thus, we fight an uphill battle when we introduce antimicrobial substances into a milieu loaded with microbes.

Jeffreys (1952) has shown that antibiotics such as griseofulvin, patulin, and mycolic acid will survive longer in organism-free soil then in a natural soil full of microbes. Walker and Smith (1952) show that *Myrothecium verrucaria*, a very widespread cellulose decomposer, can actively destroy cycloheximide (Actidione). Salicylic acid analogues have been shown to contain some systemic chemotherapeutic properties on bacterial blight of bean, but Riere (1940) has shown that *Aspergillus*, a common soil organism, can utilize salicylic acid as a source of carbon. Wright and Grove (1957) have recovered a *Pseudomonas* species from soil that can degrade griseofulvin. This bacterium increases steadily in the soil as griseofulvin is successively added to the soil. Gottlieb (1957) has published an excellent review of the microbial decomposition of toxicants. We can only conclude that chemotherapeutants destined for entry into the plant through the roots must run a very hazardous gauntlet. This inevitably suggests that screening methods for chemotherapeutants should include a "soil burial" test analogous to that used by our colleagues in wood and fabric degradation.

Even assuming that the candidate therapeutant survives the hazards of the soil and arrives alongside the root, it must pass through the epidermis of the root, migrate through the cortex, if necessary through the endodermis, and finally, into the free-flowing stream in the xylem. This formidable collection of barriers further separates the "sheep from the goats, the men from the mice." The number that survive falls fast. Here also adsorption plays its accustomed role. Of course, investigations or root entry are best done when the roots are placed in sand or in solutions or suspensions of the test substance, not in soil.

Chapman (1951) has reported on the comparative performance of S-quinolinol and *n*-octadecyltrimethylammonium pentachlorophenate. The former is almost without charge. It does pass the root barrier and go up the stem of an elm. The latter compound is highly charged, is substantive to cellulose according to Chapman (1951), and it does not move out of the roots.

The relative performance of these two therapeutants is striking. The former gets into and up the stem of elms to the point where invasion by the Dutch elm disease fungus occurs high in the tree. The latter compound is utterly ineffective on Dutch elm disease, but is reasonably active of *Fusaruim* wilt of tomato where invasion occurs through the roots. This compound stop in the roots and is effective there. The other apparently does not stop there and is ineffective there. In 1947, Anderson and Nienow demonstrated that streptomycin enters roots of soybeans. It also enters the roots of peach, cucumber, broad bean (*Vicia faba*) and tomato.

Chloramphenicol can also enter roots, whereas chloroteracycline oxytetracycline, and neomycin can not. The evidence is clear that the following other compounds can enter the plant through the root: cycloheximide, griseofulvin, thiolutin, sulfonamide analogues, 4-chloro-3, 5-dimethylphenoxyethanol, benzothiazolyl-2-thioglycolate and many others. The charge on streptomycin affects its adsorption into root tissue just as it affects adsorption onto the soil particles. Streptomycin tends to remain in broad bean roots just as *n*-octadecyltrimethyl-ammonium pentachlorophenate does, but to a lesser degree. Pramer's work also suggests that chloroamphenicol and griseofulvin are not retained in the roots; presumably, they are not heavily adsorbed there.

Cycloheximide does not seem to be heavily adsorbed onto roots tissues. Crowdy and Rudd Jones (1956), like Chapman (1951), have related the adsorption of sulfonamides by root tissues to their adsorption on cellulose. Their techniques were somewhat more sophisticated, however. They have shown that the rate of entry of sulfonamides into broad bean roots is related to their R_F values (the ratio of the distance traveled by the material to the distance traveled by the solvent) on cellulose chromatograms. This looks like a smart approach to the problem.

The ratio of entry to the R_F value on cellulose is a constant for the neutral compounds that are fat soluble, but the more water soluble the compound, the higher the proportion of aqueous phase in the chromatograph system needed to maintain the same ratio. Apparently, neutral compounds and those with high pKa values (pH where ionization is least) enter the plant as undissociated molecules, but those with low pKa values enter as ions. Of course, when the pKa value is near 7, pH is very important in dissociation and, hence, is closely associated with entry of the compound.

Bearing in mind that systemic chemotherapeutants must move through living cells of the root before entering the xylem, one is not surprised to note from the literature that the process requires energy as many permeative processes do. Griseofulvin shows an initial rapid entry into wheat roots and this is inhibited by such respiration inhibitors as nitrophenol, phenylurethane, and sodium azide. At the concentrations used, these inhibitors do not inhibit the entry of water, however, according to Crowdy et. al. (1956). If the inhibitors were applied for a prolonged period, some griseofulvin did enter the root presumably with the unihibited water of the transpiration stream.

Crowdy and Rudd Jones (1956) pursued this line of research also with the entry of sulfonamides into the broad bean plant. Some of the sulfonamides enter readily, some do not. Those that do, act similarly to griseofulvin. The initial sport in the entry is sensitive to respiration poisons, but the long time slow entry is not. The latter is related to the rate of entry of water. Presumably, one could force candidate therapeutants through the roots, but we have no information on such trials. Much research has been done on entry through wounds in the stem. Stoddard (1947) forced his test compounds into the severed upper ends of the stems of peach seedlings. He seems to have had no followers, but his methods would seem to warrant further exploration.

The uptake of chemicals by stems through bore holes has been very popular. This technique has been resorted to largely because it permits estimation of the dosage within the plant. Howard (1941) used the method for his work on diaminoazobenzene for bleeding canker of maples, and so did Zentmyer et. al. (1946) in their earlier researches on Dutch elm disease.

Radial bore holes were used at first and the compounds used were water soluble. Later, Howard and co-workers developed a method of injection that depends upon tangential bore holes arranged in an ascending spiral around the tree. The holes are filled with the compound in paste or powder form and sealed. The tangential boring assures that the chemical is accessible to the maximum number of xylem vessels and the "dry" packing assures a longer lasting supply of chemical. The method, when properly timed, seems to be successful with 4, 5-dimethylthiazolyl-2-thioglycolate for Dutch elm disease, but not for 8-quinolinol benzoate. Is a tree "cured" of disease when the development of symptoms is stopped? For example, a 25-year-old elm tree about 8 to 10 inches in diameter at breast height became naturally infected with *Graphium ulmi* as proved by tissue culture from wilted terminal branches. Immediate chemical treatment upon sighting the symptoms checked further wilt. Subsequently, the dead branches were removed and the tree appeared healthy (symptomless) for 2 years until another attack developed. This was confirmed as due to *Graphium ulmi* by laboratory diagnosis.

Again prompt treatment by the bore-hole method resulted in checking the disease and it has remained "healthy" for 3 years. In another case, an elm was proved to be infected with *Graphium ulmi* by isolation of the fungus from branches bearing wilted leaves. It was promptly treated by soil impregnation of the rhizosphere and a further development of symptoms was stopped, The next two seasons, normal foliage appeared on undamaged branches and the killed ones were cut off. However, discolored streaks in the xylem, formed during the year of active disease, were separated from the cambium by annual layers of new wood. This discolored wood yielded pure cultures of the pathogen. Thus, it would appear that a sometime pathogenic fungus can continue to exist in a quiescent state in a host plant without causing injury.

There may be a "walling-off" effect brought about by the interposition of cells having thicker walls or unfavorable chemical composition, or the metabolic environment is not favorable for pathogenesis. To test this, ax cuts were made through the bark and into the discolored area harboring the "quiescent" pathogen, with the result that in some cases symptoms again developed. Beckman (1958) now believes that arresting Dutch elm disease by 4, 5-dimethylthiazolyl-2-thioglycolate treatment creates an unfavorable physiological environment, rather than a mechanical barrier for walling-off the fungus.

Sometimes severed stems are used in exploratory work on chemotherapeutants especially in studies on host tolerance. Robison et. al. (1954) have shown that antibiotics enter more readily through cut stems than through roots, but this, of course, is not unexpected. Some work has been done on entry through intact stems. This is a technique first used by growth hormone investigators. Mitchell et. al.

(1952) applied streptomycin to bean stems by mixing it with lanolin and a nonionic detergent. Streptomycin will also enter the intact stems of tobacco. Polychlorobenzoic acid will readily enter when painted on the bark of young elms and affect the development of the Dutch elm disease. If we could find out or tailor-make chemotherapeutants that enter readily through intact foliage, we would be a long way toward a practical solution to the entry problem.

If the compound enters the foliage, it is very near to the site of action of all above-ground diseases. It need not be translocated very far, and it can be easily and cheaply applied as a spray or, perhaps, as a dust. Many compounds do enter through intact foliage; some directly through the cuticle and others through natural openings. This problem also has been pioneered by the growth-hormone investigators. The problems are similar to those encountered in root entry. The compound must pass through the epidermis, through the parenchyma tissue, and into the xylem or the phloem. The problem is further complicated by the presence of the waxy cuticle which is pretty inert. It is eased by the absence of soil and its complications and it eased by the presence of stomata and hydathodes. Curtis (1943) has shown that compounds can enter through the hydathodes where guttation drops form.

Crafts (1918) has contrasted the types of compounds that might enter through foliage or through roots. He feels that compounds that enter the leaf should be somewhat less polar and more lipid-soluble than those enter roots. A mixture of aluminum sulfate, kaolin, and calcium carbonate, (Cuneo Mixture) administered either to the foliage or in proximity to the roots of trees or herbaceous plants, is claimed to act as a systemic fungicide for several diseases. Mosca believes that all polar (+ charged) compounds of metal derivatives conductive of electricity have fungicidal properties. Their fungitoxicity is proportional to the degree of dissociation. When a polar substance is dissolved in water, it decomposes into atoms furnished with electric charges (free ions)—the ionization of polar substances is spontaneous. According to Mosca (1952), Al^{+++} in the ion state acts as a "systemic fungicide with a universal action", because the aluminum ions are conveyed by the protein substances throughout the plant.

Of course, it is now common knowledge that phenoxyacetic acid analogues penetrate foliage very easily. It is interesting to note that sodium benzothiazolyl-2-thioglycolate penetrates elm foliage and has definite therapeutic effects on Dutch elm disease. This compound has an -S—CH_2COOH, which is the sulfur analogue of the —O—CH_2COOH side group of 2, 4-D. Similarly, van der Kerk's (1956) new systemic fungicide has precisely the same group attached to it. This compound is disodium ethylene bisdithiocarbamyl-dithioglycolate. It is also interesting that captan has a similar tail (—S—CCl_3) attached to the imide grouping. The -S—CCl_3 has the important similarity of -S—CH_2COOH in that the electronegative groups C = O and Cl_3 are separated by one carbon from the sulfur.

Captan does not enter intact foliage as easily as sodium benzothiazolyl-2-thioglycolate or 2, 4-D, but it does seem to possess local chemotherapeutic properties as Stoddard (1954), and Rich (1956) have shown for cucumber scab. It would appear that -S—CH_2COOH is a better "shaped charge" than —S—CCl_3.

These results suggest that, perhaps, permeation and, perhaps, translocation are encouraged by R—S—R' or its oxygen analogue. When one surveys some of the known synthetic chemotherapeutants. The R seems to vary but apparently the R' should contain an electronegative group. From these structures, one can see others that perhaps should be tried. Nirit, the protective fungicide, is 2, 4-dinitropheylthiocyanate. Here the cyanate group provides the electronegativity. Perhaps, it would be therapeutic. Perhaps, substitution of chlorines for the nitro groups would make it a close analogue of 2, 4-D and, perhaps, improve its activity.

Antibiotics also enter uninjured plant leaves. Davis and Rothrock (1956) demonstrated that griseofulvin applied to one surface of tomato leaves controlled *Alternaria solani* applied to the other. Similarly, Dye (1956) applied streptomycin to the lower surface of peach leaves and protected them against *Pseudomonas syringae* applied to the upper surface. A few data are available on the weather factors that influence entry through intact foliage. The longer the applied material remains damp the longer the time available for entry. Thus, highly humidity delays the drying and promotes entry as Dye (1956) has shown for streptomycin. This principle has been exploited by Gray (1955) to improve the entry of streptomycin. He used glycerine in the water drops of streptomycin to delay drying. It improved the entry.

Goodman (1954) has used Cellosolve and Carbowax 4000 for the purpose of increasing the entry of streptomycin and oxytetracycline for fire blight control on apple and pear. Glycerine sometimes does not improve entry as Shaw et. al. (1957) have shown for tobacco foliage and Rich (1956) for cucumber foliage.

2. *Translocation.* If we are to attain systemic chemotherapy, we must attain excellent translocation from the point of entry to the point of need. If we are concerned with leaf spots, we must obtain long distance translocation upward if we use root application. If we are concerned with root rots, we must obtain long distance translocation downward if we use foliage application. Conversely, foliage application is more likely to succeed for foliage diseases and root application for root diseases. Vascular diseases and systemic viruses are hard to reach from either end. We in plant pathology are more hampered than our medical colleagues by translocation problems. An antibiotic injected into an artery or into a vein will appear in all parts of an animal in a fraction of a minute. The plant, unfortunately, has no such circulation system. And what circulation system it has occurs in microscopic units. We have no tubes as big as the arm vein, for example. We cannot inject very well the whole body from a single point.

The upward system is different from the downward. By and large upward translocation seems relatively simple. Many compounds applied to the roots will appear pretty well all over the tops, but compounds applied to the tops often do not travel downward at all. One day this problem will have to be solved. Delay in its solution will delay the success of chemotherapy of plant diseases. The assay of translocation is seldom simple. A few antibiotics produce characteristic biological effects. For example, griseofulvin produces a characteristic curious curling of the germ tubes of *Botrytis allii*. Its presence can, thus, be detected and has been so used by Brian (1952) and his associates.

The bacteriologists have developed strains of *Escherichia coli* that must have streptomycin or they die. If streptomycin-treated plant tissue supports their growth one can be reasonably sure that streptomycin is there. Mitchell et. al. (1953) have studied streptomycin uptake and translocation with this dodge. Van Raalte (1952) has devised a partial answer to translocation. He applies the test compound to a cut end of a short piece of potato stem standing on an agar plate seeded with bacteria or other organisms sensitive to the test compound. A zone of inhibition indicates translocation.

Phelps et. al. (1957) have shown that cycloheximide, cycloheximide acetate, and oligomycin travel in the xylem if injected into the stems of oak trees. Robison et. al. (1954) have found that antibiotics vary greatly in xylem movement even if they enter unhindered through cut stems. Presumably, this is the ancient old bugaboo of adsorption to the cellulose. The tetracyclones move upward much more rapidly than streptomycin and neomycin. We have already learned, however, that the streptomycin molecule is charged and that it is readily adsorbed. We have seen that K_2HPO_4 will elute streptomycin form clays. The same authors also have shown that K_2HPO_4 will keep streptomycin moving in the xylem elements of *Pyracantha* cuttings.

Streptomycin appears to move in the xylem more rapidly in the peach than in some plants because the concentration in the leaves rises above that in the roots. In the broad bean streptomycin concentrates more in the roots than in the leaves. Griseofulvin and chloramphenicol move steadily out of the treated broad bean roots and accumulate in the foliage. Griseofulvin acts in tomato, according to data of Crowdy (1957), like streptomycin does in the broad bean. There is a gradation form the roots right out to the highest leaves. It must be bound by the first sites and moves further only as the lower sites become saturated.

There seems no doubt that many of these compounds tend to move in the xylem because some of them show up in the guttation drops that exude from the hydathodes, i.e., cycloheximide, and griseofulvin. Other evidence is that entry of some antibiotics over the long pull or in presence of respiration inhibitors is proportional to the amount of water lost in transpiration. On the other hand, Hidako and Murano (1956) concluded that if streptomycin is applied to tomato stems it moves in the pith and phloem not in the xylem. Napier et. al. (1956) applied streptomycin to the simple primary leaves of bean plants and observed an effect on halo blight as far away as the fourth trifoliate leaf. Crowdy and Pramer (1955) summarized data on translocation. They concluded that neutral or acidic compounds such as griseofulvin move readily in the plant, that basic compounds such as streptomycin and amphoteric compounds such as the tetracyclines give different results with different plants. Our ignorance of the field is profound.

3. *Host Action on the Therapeutant.* No host will sit idly by when it is treated with a therapeutant. It may detoxicate a therapeutant, enhance its effect, or possibly even excrete it through the leaves or roots. Hilborn (1953) has presented a tantalizing abstract showing how different hosts affect chemotherapeutants. We regret that he has not elaborated it. He shows, for example, that chloramphenicol reduces verticillial wilt of potato, but not of tomato.

(a) *Detoxication in the Host.* We have already seen how soil microbes may destroy a chemotherapeutant. Likewise, the living host also may degrade a chemotherapeutant and render it impotent. This is a very well known phenomenon in human medicine and much is known of the mechanisms of degradation. In plant chemotherapy, however, about all we know is that some compounds must be degraded because their effect is lost. 8-Quinolinol derivatives are evanescent in the plant. The sulfate lasts a few days, the benzoate a few days longer. We do not know that happens to them, but perhaps they are degraded.

The term "half-life" is common these days to quantify the rate of loss. One may, thus, rate chemotherapeutants in terms of half-life. Some of the systemic phosphate insecticides have a half-life that can be expressed in weeks, but most microbial chemotherapeutants have half-lives measured in days, perhaps even in hours.

Brian (1952), for example, showed that griseofulvin has a half-life of only a few days. Crowdy (1957) has determined this number to be about 4 days. Perscot et. al. (1956) have shown that cycloheximide has a half-life of bout 24 hours in the cherry fruit but much longer in cherry foliage. According to data of Wallen and Millar (1957), cycloheximide may have a half-life in wheat foliage of 2 or 3 weeks. In the human body, acetylation is a very common device for detoxicating drugs. Esterases are common among most living organisms. One would expect that higher plants would be able to make esters out of poisonous substances. Unfortunately, only a few have searched for such reactions. Rudd Jones and Wignal (1955) have shown, however, that the broad bean can acetylate sulfanilamide.

Undoubtedly, many more cases of such reactions await discovery. We shall return to this reaction below in discussing possible means of producing effective chemotherapeutants. Other devices are undoubtedly available to host plants in their fight against a foreign chemical, the introduced therapeutant. Gottlieb (1957) has discussed how microbes can detoxicate toxicants. Some, perhaps all, of these mechanisms are available to the host. He mentions hydrolyses, esterification, chelation, direct reaction with metabolites, oxidation and reduction, and displacement effects. We need much more research in this field. Sanwal (1956) has shown that fusaric acid is decarboxylated by the tomato plant and transformed into a nontoxic substance.

(b) *Activation by the Host*. The host can react with a compound to enhance its effect, just as it can react with a compound and reduce its toxic effects. Martin (1950) showed fairly early that the systemic insecticide bis-(bisdinethylaminophosphonous)-anhydride is more toxic after it has been acted on by the host than before. Similarly, the effectiveness of 4-chloro-3, 5-dimethylphenoxyethanol to reduce fusarial wilt of tomato improves with time. This type of evidence is not conclusive proof that the compound actually changes. The improved effectiveness with time could be due to production of a more effective compound or to an increase in the resistance of the host.

The phenoxyethanol probably is oxidized in the plant to phenoxyacetic acid. Since, however, the effects of phenoxyacetic acid also increase with time, one is more inclined to suggest as did Davis and Dimond (1953) that the resistance of the host is increased as well. Gray (1958) has evidence that tobacco tissue converts streptomycin amine and streptomycin oxime to a more toxic compound, probably into streptomycin itself or dihydrostreptomycin. Similarly, Lemin and Magee (1957) have found that cycloheximide acetate is deesterified by the host to cyclohemixide itself. Rombouts and Sijpesteijn (1958) suggest that the cucumber and broad bean may convert the nonfungitoxic carboxy methyl derivative or pyridine-2-thiol-*N*-oxide back to the fungitoxic thiol compound.

(c) *Excretion from the Host*. Our medical colleagues have made great strides in learning how chemotherapeutants are lost from the host by excretion. Indubitably, plants also excrete chemotherapeutants but our knowledge of the subject is meager indeed. Presumably, the compounds can be excreted through the leaves or through the roots. In the leaf, the most obvious excretory organ is the hydathode. Some chemotherapeutants have been recovered from guttation water. Notable among these are griseofulvin and cyclohemixide. Gopalkrishnan and Jump (1952) were not able to recover thiolutin in the guttation water.

The idea that roots excrete all manner of compounds has only recently been given detailed consideration. We must have more knowledge about root excretion of compounds of we are to combat root rot diseases by spraying the foliage. Halleck and Cochrane (1950) and Zentmyer (1954) have made a start to search for compounds that could be applied to the foliage, translocated to the roots, and excreted into the soil. This would appear to be a promising lead to follow.

Modes of Chemotherapeutic Action

Having found an effective therapeutant; having preserved it from degradation by soil, microbes, and host; having found that it is translocated to the infection court; having saved it from excessive excretion, what do we know about its mechanisms of action? Research so far suggests three major mechanisms of action: vivotoxins are destroyed; the pathogen is killed by direct action; or the resistance of the host is enhanced. These we shall now discuss.

Inactivating Vivotoxins

Plant pathogens excrete sewage products into the medium where they grow. Sometimes these are called "staling products", because they inhibit the very fungus that excretes them. Plant pathologists, seeing such effects in fungus cultures, have suggested for a half century or more that such products must also be produced by the fungus in the invaded plant and must participate in the pathogenesis of disease. In such cases the substances are called toxins. This has always been an intriguing hypothesis. It is discussed in all of its implications by Ludwig.

Dimond and Waggoner (1953) insisted that the alleged substances must meet Koch's rules of proof, if they are to be implicated as casual factors in disease. For one thing, they must be recovered from the diseased host. It is not enough that they be found in laboratory media. Those that can be recovered from the hosts were labeled vivotoxins. Since vivotoxins probably do occur in the chain of causality, one is intrigued by the possibility that pathogenesis can be reduced by neutralizing such toxins.

Howard (1941) first proposed such a possibility to account for his alleviation of the symptoms of bleeding canker of maple with diamino-azobenzene. Zentmyer (1942) and Stoddard (1946) offered a similar explanation for the action of 8-quinolinol in alleviating symptoms of Dutch elm disease. In neither case was a vivotoxin proved. Perhaps the best example of a vivotoxin is that of the wildfire disease of tobacco as investigated by Braun (1950) and by Woolley et. al. (1952). The vivotoxin is a complicated amino acid analogue of methionine and it can be antidoted chemotherapeutically with methionine. A very new comer as a vivotoxin is fusaric acid which clearly is involved in the pathogenesis of fusarial wilt. It is an analogue of pyridine.

Subramanian (1956) showed that its effects could be antiodoted with 8-quinolinol thus putting props under Horsfall and Zentmyer (1942) who had said that 8-quinolinol seemed to mitigate the symptoms of Dutch elm disease by antidoting a toxin. However entrancing the vivotoxins may be in the genesis of disease, their treatment by chemotherapy may be a delusion. As long as the pathogen remains in the tissue it will presumably continue to excrete the vivotoxin. That means that any antidoting chemical must be maintained continuously in the system. This will be difficult. One is in the position of Alice in Wonderland. He must run as hard as he can to stay even.

Horsfall (1956) says that he had hopes that perhaps the destructive action of the toxin could be kept under control long enough to enable the host to throw off the invader by other means. As yet, this still remains only a hope. The future looks dark for successful chemotherapy by toxic control alone.

Direct Action on the Pathogen

If the microbe is a primary causal factor in disease, surely we must kill it out if we are to save the host. Success in so doing is very elusive, however. For many years after the modern reactivation of research on chemotherapy, the number of cases of direct killing were scarce indeed. Before proceeding to a discussion of such successes as we have had, we should consider direct killing of the pathogen by substances found in nature. These, perhaps, will be useful to suggest further research on synthetic substances.

1. *Natural therapeutants.* Plants display a wide variety of cases where resistance to disease seems chemical in nature. The role of phenols and tannins in natural resistance has been discussed for years. The classic example is protocatechuic acid which has been shown by Walker et. al. (1929) to be responsible for the resistance of red onions to smudge. Natural therapeutic agents, such as phenolic compounds, have been found in trees by Rennerfelt (1945), and Erdtman (1949).

Rich and Horsfall (1954) have called attention to the fact that quinones are generally more fungitoxic than phenols. Schaal and Johnson (1955), investigating the role of chlorogenic acid in resistance of potato to scab, suggest that the phenol is oxidized to a quinone before maximum resistance is attained. Presumably, this oxidation occurs when the potato tissue is damaged upon invasion. The phenol situation is interesting in fusarial wilt of tomato.

One of the outstanding characteristics of their disease is the darkening of the vascular bundles. Davis et. al. (1953) asked themselves, what is the color and how does it originate? After studying the problem, they decided from their date that the pigment is melanoid, that it derives from a phenol which is oxidized to a quinone, and polymerized to a pigment. They decided that the phenol comes from aphenolic glycoside in the host, that the hydorolytic enzyme comes from the invading fungus, and that the phenoloxidase comes from the host. In brief, the course of events in blackening of the vascular bundle is that a phenol glucoside of the host is hydrolyzed by a fungus β-glucosidase, that the host enzymes oxidize the phenol to a quinone and convert the quinone to a pigment.

Presumably the same situation occurs in other wilt diseases, such as Dutch elm disease, where the vascular elements are discolored.

Here, then, is a parallel situation to the potato scab. In potato scab the chlorogenic acid phenol that is liberated is somewhat fungicidal and its quinoidal analogue more so. Presumably, then, the phenols and the quinones that are precursors of the pigments in vascular diseases are fungitoxic also. The question is, do they discourage invasion? Perhaps the Dutch elm disease might be considered first. We know that the so-called black strain of the Dutch elm disease fungus is a much more pathogenic strain than the light colored strain. Presumably, this can be explained by the results of Rich and Horsfall (1954). They found that *Monilinia fructicola*, a hyaline fungus, is poisoned by phenols and quinones while *Stemphylium sarcinaeforme*, a black fungus generally is not.

The black fungus is able to detoxicate phenols and quinones by converting them to bland black pigments, but the hyaline fungus cannot. Possibly, the hyaline strain of the Dutch elm disease fungus is unable to detoxicate the phenols or quinones formed upon invasion, and is thus poisoned. The black strain can detoxicate them, is not poisoned, and thus produces more damage than the hyaline strain. Similarly, Taylor and Decker (1947) report that the strains of the potato scab fungus that form black pigments are more pathogenic than hyaline strain. One wonders how the blackening of the tomato xylem fits into the fusarial wilt picture. Apparently the fungus, being colored, can tolerate the phenols or the quinones that are precursors of the pigment. Perhaps the fungus can detoxicate the compounds as *Stemphylium sarcinaeforme* does. One observation is interesting. Gothoskar et. al. (1955) have shown that respiratory inhibitors such as 2, 4-dinitrophenol, sodium flouride, thiourea, and sodium diethyldithiocarbamate break down the resistance of the Jefferson tomato to fusarial wilt. The last two named are active inhibitors of phenoloxidases. Perhaps these two, at least, prevent the formation of a toxic quinone and thus destroy natural resistance.

The case of verticillial wilt of tomato is interesting in this context. Here is a vascular disease in which the vascular system is only weakly discolored, if at all. The quinone or the melanin reaction must be blocked somewhere. This was mysterious until Caroselli (1955) discovered that *Verticillium* seems to secretes thiourea, at least in culture. If, indeed, it secretes thiourea in the tomato, it could inhibit the polyphenolase system and, thus, prevent the blackening. It probably also prevents the formation of the toxic quinone by the host and thus protects itself from damage. Stated in Gothoskar's terms, the thiourea secreted by the *Verticillium* fungus breaks down the host resistance. The paper of Voros et. al. (1957) fits into an interesting niche here. Streptomycin is not fungitoxic but as a chemotherapeutant, it seems to increase the resistance of the potato to *Phytophthora infestans*. These authors give evidence that streptomycin activates the phenol oxidase system of the potato so that it produces more of the natural substances (possibly quinones) that are responsible for natural resistance.

One wonders if it would not then increase the resistance of the potato to scab. Certainly much more needs to be leaned about the polyphenol system. Apparently we can increase susceptibility by inhibiting it, decrease susceptibility by stimulating it. This still leaves hanging the case of the tomato wilt disease. The quinone reaction does not seem to affect the situation in the normal infections. The Bonny Best variety is very susceptible, but the vascular bundles turn very dark. Here is an anomaly. The phenol → quinone → pigment reaction exists, but the fungus is not poisoned. This cannot be a thiourea case, either, because this would prevent coloration.

We can marshal what evidence we have. Davis et. al. (1953) showed that *Fusarium oxysporum f. lycopersici* can hydrolyze phenol glycosides and use them as sources of carbon. Presumably, this suggests that the phenol formed in the tomato stem is not toxic to the fungus. The phenol apparently is oxidized by the host to a quinone. We do not know whether it is toxic to *Fusarium* or not. From their colors, we know that *Fusaria* can produce many pigments. Probably, *F. lycopersici* can detoxicate a quinone by forming a pigment, as *Stemphylium sarcinaeforme* does. The results of Dimond and his colleagues on phenoxy and naphthoxy acetic acid analogues seem to find a place somewhere in this area of the reasoning.

Essentially, all of these compounds inhibit the blackening reaction in tomato stems that have been inoculated with *Fusarium*. On this basis, the compounds have been rated as chemotherapeutic. They do indeed knock out the commonest symptom of disease. We are unable to discover any other work on the effect of these compounds to inhibit the polyphenolase blackening reaction of higher plants. Several authors, however, report that 2, 4-D inhibits pigmentation in dark-colored fungi, as for example in *Helminthosporum victoriae*. It seems reasonably safe to assume then that these compounds do inhibit pigment formation. Do they inhibit quinone formation as well? We have no evidence here. The quinone may or may not be important, because the fungus does seem to be able to cope with it in nature.

One wonders what effect the phenoxy compounds would have on the resistant Jefferson variety of tomato. One wonders if the effect would be similar to that of the inhibitors of polyphenolase as reported by Gothoskar et. al. (1955). Whether they act on the quinone or not, the fungus is not killed by them because when medication is withdrawn for a few days, the tomato dies from its infection. These substances are phenol analogues. Possibly they inhibit blackening by acting as phenolic antimetabolites. This brings to mind another interesting result that has occurred in Dimond's tests. 2, 4-Dichlorophenoxypopionic acid is not as effective in reducing the blackening reaction as 2, 4-dichlorophenoxyacetic acid. A similar result has been shown on chocolate spot of bean by Crowdy and Wain (1950). This disease also is read in terms of the blackening reaction. The propionic acid derivative does not reduce the blackening. The acetic acid derivative does.

According to Synerholm and Zimmerman (1947), the propionic acid analogue is degraded to a phenol by the host, the acetic acid analogue is not. Thus, the former would add to the blackening, if anything, the latter would not. What, then, about other symptoms of disease? What about wilting? Generally, the growth substances reduce wilting too. However, Dimond has told us verbally of a compound which prevents the blackening of the vascular bundles but which permits 73% as much wilting as the check. This compound is $2,4—Cl_2C_6H_5—O—(CH_2\text{-}CH_2\text{-}CH_2\text{-}O)_2\text{-}CH_2CH_2OH$. A homologue with four ethoxy groups in the side chain gave similar results. It is obvious that these tow compounds can eliminate one symptom of tomato wilt, but not others.

2. *Synthetic Therapeutants*. Until the modern revival of interest in synthetic therapeutants, most of the cases of successful cure of plant disease were limited to local or topical chemotherapy. Apple scab lesions could be cured with lime sulfur and organic mercurials. Certain cereal smuts could be cured by treating the infected seeds with organic mercurials, but we probably had no translocatable "systemic fungicide."

By our definition, a systemic funicide is one that distributes itself systemically through the plant and kills the pathogen by direct toxic action. Crowdy and Wain (1950) spoke of the systemic fungicidal activity of analogues of phenoxyacetic acid. These compounds are only weakly fungitoxic, however. This led Horsfall and Dimond (1951) to suggest that they probably act to make the host resistant rather than to kill out the pathogen. Later researches seem to bear this out. Crowdy and Davies (1952) showed that phenyl mercury acetate is translocated from the roots to leaves of bean plants and inhibits the chocolate spot disease.

Antibiotics probably showed the first really good case of a systemic therapeutant that could indeed kill out the invader. Now we know that cycloheximide possesses such properties for heat rust and cherry leaf spot, cycloheximide semicarbazone for *Coccomyces hiemalis* in sour cherry pyridine-2-thiol-*N* -oxide for cucumber scab, 4-nitrosopyrazole for *Alternaria* on tomato, and griseofulvin for several diseases. The antifungal polypeptide, "Phytoactin", has markedly reduced infection with oak wilt when applied in solution through trunk cuts to northern pin oaks. In University of Wisconsin experiments on oaks, 3 to 10 inches in diameter and 20 to 40 feet in height, oligonycin yielded an average of 13% healthy trees, Actidione 20% its acetate derivative 30% for three seasons, and Phytoactin 40% after two seasons. By comparison, 95% of the untreated, inoculated check trees had died. A group of inoculated greenhouse and nursery oaks in Ohio tests were not benefited. A few reports suggest that viruses can be inactivated in plants. Stoddard (1947) reported curing peach seedlings of X-disease by injecting the outer ends of the stems with calcium chloride and zinc sulfate.

The zinc effect has by now been investigated for numerous virus diseases. It seems to be effective for some such as carnation mosaic but not for others such as tobacco mosaic virus on *Nicotiana glutinosa*. Respiration inhibitors seem to inhibit virus multiplication. Similarly, some antimetabolites seem to be effective for viruses, as for instance sulfanilande, thiouracil, and guanazolo. Therapeutic action of sulfonamide and related sulfa compounds is currently believed to be by interference with enzymes necessary for growth. They disturb the utilization of para-aminobenzoic acid and, hence, inhibit the action of certain coenzymes essential for bacterial and actinomycete growth. A possibility for curing systemic bacterial diseases is the internal application of translocatable sulfones capable of breaking down *in vivo* into diamino- diphenyl sulfone which exercises antibiotic action similar to the sulfonamides.

3. *Possibilities for a Tailor-Made Therapeutant*. Perhaps something can be derived from our experience to date to suggest possibilities to make better therapeutants. As we have seen, phenols and quinones seem to be involved in natural host resistance. It seems probable that our knowledge of these compounds will increase. The evidence is fairly clear from Allen's review and from the work of Davis et. al. (1953) that in the very act of invasion phenols may be freed from natural nontoxic compounds such as phenolic glycosides. This suggests that we feed such detoxicated phenols to the host. They should not be phytotoxic.

They would lie in wait for an invading microbe which would free them from their bound form. Thus, the microbe would poison itself at the point of invasion. In effect, this would be a case of hypersensitivity.

The work of Byrde and Woodcock (1952) suggests a device that would work on this principle. 2, 4-Dichloronaphthoquinone is a powerful fungicide used commercially as a protectant. Byrde and Woodcock reduce it to the dihydroxynaphthalene form and then acetylate the two phenolic groups. This procedure renders the compound nontoxic to apple foliage when applied to the surface. The apple scab fungus, *Venturia inaequalis*, can deacetylate the compound by means of a fungus esterase.

According to Byrde and Woodcock, the dihydroxynaphthalene thus liberated is toxic. Horsfall (1956) suggested that the compound is probably oxidized by the fungus back to the still more toxic naphthoquinone. If now, the diacetoxy derivative were used as a chemotherapeutant, it should be the answer to our prayers. The invading fungus should free the phenolic form, the host should convert it to a quinone. It cannot be converted to a pigment because the chlorine blocks the important position. This, it should be an ideal chemo-therapeutant. The concept is too simple. Upon trial we find that the host esterases can liberate the phenol and oxidize it as readily as those of the fungus. The host dies by its own hand even before the pathogen arrives. It works on the surface of an apple leaf, but not inside the tissue.

Other acetylated phenols have followed this one down the drain. Esterases for acetic acid deriatives seem to be too common in nature. Perhaps, we need to investigate other acids, such as benzoic. The linkage here is harder to break according to Byrde and Woodcock (1956), but on the other hand the fungus esterases may not break it either. An interesting ester is the thioglycolic ester of ethylene-bisdithiocarbamic acid as proposed by van der Kerk (1956) as a systemic chemotherapeutant. According to van der Kerk the host deesterifies this linkage and recovers the nabam itself. Perhaps phosphate or borate esters would work better. The phosphates have solved the problem for systemic insecticides. Perhaps amide linkages are better sources of detoxication. Whether fungus amidases are more common or more specific than host amidases, can only be settled by empirical attack. In any case, this appears to be a lucrative source of possible chemotherapeutants that might act differentially on host and pathogen.

Development of Resistance to Chemotherapeutants

As soon as one speaks of chemically killing a pathogen, one cannot avoid the possible development of resistance by the pathogen. This problem has plagued our medical colleagues almost from the day of the development of the first successful chemotherapeutant. Fortunately, it has not been a serious problem for the plant pathologist, perhaps because we do not yet have a chemotherapeutant in widescale commercial use. Acquired resistance to chemotherapeutants has been reported for bacteria. Mitchell et al. (1952) report that *Xanthomonas phaseoli* causing bean blight develops resistance to streptomycin. Strangely enough the halo blight bacterium on beans, *Pseudomonas phaseolicola*, does not develop resistance so readily. *Erwinia amylovora* on apple and *Xanthomonas vesicatoria* on pepper also rapidly develop resistance to streptomycin. English and van Halsema (1954) have shown that plant bacteria develop resistance less rapidly if more than one antibiotic is used.

Increasing Host Resistance

Increasing natural resistance to development of the pathogen by chemical modification of the host cell is possible. Weintraub et. al. (1952) suggest that the antiviral effect of Stoddard's (1947) zinc sulfate is probably a case of increasing host resistance. Chemotherapy may also be used to induce extreme susceptibility or hypersensitivity. This phenomenon of hypersensitivity occurs naturally between wheats and the rust fungus, potatoes and the late blight fungus, etc.

1. *Effect of Auxins*. Plant growth regulators give striking effects in the amelioration of some diseases. These effects were discovered independently and reported first by Crowdy and Wain (1950) for the chocolate spot of beans and then by Dimond and Chapman (1951) for fusarial wilt of tomatoes. More recently they have been tested by Beckman (1958) for Dutch elm disease. These compounds were selected originally for testing because they were known to be translocated.

Dimond and his group continued their researches on the auxins and auxin-like substances and found that almost any auxin that acts on the tomato reduces the fusarial wilt syndrome. The striking feature, however, is that the funitoxicity of auxins is extremely low, so low in fact that it could not possibly account for the reduction in disease. This discovery led to the conclusion that the auxins somehow make the plant resistant to disease. The nature of this mechanism is still vague, but some progress is being made. The final solution must await information on just what the auxins do and can do to the plant.

2. *Relation to Sugars*. In the meantime, a few statements can be made. For one thing, auxins exert a strong influence on the sugar levels in plants and Horsfall (1956) hazarded a conjecture that the

sugar levels are correlated with disease. This phase of the matter has been pursued in considerably more detail by Horsfall and Dimond (1957). The effect of auxins on sugar levels varies with the time that elapses after treatment. Usually the sugar rises at first and then falls. Maleic hydrazide affects the phloem and reduces translocation of sugars from the leaves. Hence, it usually raises the sugar in the leaves, reduces it in stems and roots. Diseases may be high sugar diseases or low sugar diseases, that is, encouraged by high sugar or discouraged by high sugar. Rusts, powdery mildews, and chocolate spot of broad bean are sugar diseases. They should be reduced by 2, 4-D and increased by maleic hydrazide, and they are.

On the other hand, helminthosporial and alternarial leaf spots and anthracnoses seem to be low sugar diseases. 2, 4-D increases them. Fusarial wilt seems also to be a low sugar disease; 2, 4-D decreases it at first when stem sugars are high, but the effect wears off as sugar levels fall. Maleic hydrazide, on the other hand, it increases fusarial wilt, presumably because maleic hydrazide reduces stem sugars. We have difficulty in believing that the sugar effects can stand alone. Possibly the sugar effect is direct on the high sugar diseases. In this case sugar reduction could be directly related to disease reduction. The effect on low sugar diseases is less easy to follow. It seems probable that the sugar is a marker for something else.

The pathogens that induce low sugar disease are sugar feeders just like the rusts and powdery mildews. It seems likely that low sugar must be correlated with low something else, and that the low something else is related to low resistance. We must search for the "something else" that might account for resistance to *Fusarium* and to the leaf spots mentioned. Growth hormones generally increase the resistance of tomato plants to fusarial wilt. For a series of related growth regulants, Corden and Dimond (1959) studied the types of growth hormone activity with which chemotherapeutic activity is best correlated. They showed that the ability of compounds to inhibit root elongation is well correlated. The relation of this hormone function to the nature of pectin composition of the host plant is suggested by this study. Various chemicals of therapeutic promise against the Dutch elm disease are antagonistic to indoleacetic acid as judged by the pea epicotyl elongation test. A good correlation was found between the inhibition of spring wood development and suppression of symptoms. And thus we end our story. This is where our ideas of "Plant Disease Therapy" stand today. But, tomorrow—who knows?

8

Biological Control

It has been estimated that 13 % of the potential world crop yield is lost to pests. Plant pests range from nematodes to birds and mammals, but insect pests cause a major proportion of the total pest damage to crops. This chapter will therefore focus upon biotechnological solutions to the problem of insect pest damage to crops.

Far from reducing pest damage, modern agricultural practices have, if anything, exacerbated the problem. One of the contributory factors to this exacerbation has been the breeding out of endogenous pesticidal traits. Some defensive characters may have been lost accidentally during the selection process for other properties, but pesticidal traits may also have been removed 'deliberately' because they affect other characteristics like yield and quality. For whatever reason, the fact remains that many current elite cultivars have less natural resistance to pests than did their predecessors. This problem is compounded by the practice of growing monocultures, since the growing of a single crop over a wide area in repeated years encourages the build-up of pests. These negative effects of modern agriculture on pest damage have encouraged an increasing reliance on chemical pesticides, but this has tended to drive a cycle in which each new pesticide just keeps pace with the appearance of resistance to the previous pesticide in the insect population. Against this background, the possibility of developing insect-resistant crops that could help to reduce insect damage, whilst reducing the reliance on chemical insecticides, has a number of attractions. This chapter will consider the scientific strategies for achieving pest resistance before returning to discuss the potential benefits and drawbacks of GM approaches.

The Nature and Scale of Insect Pest Damage to Crops

Before examining GM strategies for developing insect resistance, it is useful to consider some of the characteristics of the insects causing the damage. The first point to make is that, whilst some adult insects feed off plants and can damage crops (think of a plague of locusts), most of the problems are caused by insect larvae. The major classes of insect that cause crop damage are the orders Lepidoptera (butterflies and moths), Diptera (flies and mosquitoes), Orthoptera (grasshoppers, crickets), Homoptera (aphids) and Coleoptera (beetles).

GM Strategies for Insect Resistance: The *Bacillus thuringiensis* Approach

In this chapter, two approaches will be compared: first, the use of bacterial insecticidal genes to provide protection from pest damage; and, second, the potential for using endogenous plant protection mechanisms (a 'Copy Nature' approach). By far the most widespread example of the first approach is the use of the *cry* endotoxin genes from *Bacillus thuringiensis*.

B. thuringiensis was discovered by Ishiwaki in 1901 in diseased silkworms, and was subsequently classified and named after its isolation from the gut of diseased flour moth larvae in Thuringberg, by Ernst Berliner. The adverse effect of the bacteria on the insect larvae was subsequently identified as arising from a number of toxins produced by the bacteria. The bacterium produces an insecticidal crystal protein (ICP) which forms inclusion bodies of regular bipyramidal or cuboidal crystals during sporulation. ICPs are one of several classes of endotoxins produced by the sporulating bacteria, hence they were originally classified as δ-endotoxins, to distinguish them from other classes of α-, β- and γ-endotoxins.

The structure of the *cry* genes and their d-endotoxin products have been well characterised. The *cry* genes are carried on plasmids and belong to a superfamily of related genes. This classification was introduced in 1998 and supersedes previous nomenclature that may still appear in the literature. The sequence comparison indicates a large number of distinct families *(cry1-cry40* at the time of publication) and within each family there may be three further levels or ranks of subfamily. Hence, *cry1Aa1* is very closely related to the other *cry1Aa* genes, which are all closely related to the other *cry1A* genes, which all form part of the *cry1* family. Table 8.1 shows that the strains of *B. thuringiensis* produce a wide range of different crystal proteins.

Thus, it is not just a case of each strain containing one specific *cry* gene encoding one particular crystal.

Table 8.1. The range of insecticidal crystal proteins in individual *Bacillus thuringiensis* strains

B.t. subpecies and strains	*Crystal protein*
aizawai	Cry1Aa, Cry1Ab, Cry1Ad, Cry1Ca, Cry1Da, Cry1Eb, Cry1Fa, Cry9Ea, Cry39Aa, Cry40Aa
entomocidus	Cry1Aa, Cry1Ba, Cry1Ca, Cry1Ib
galleriae	Cry1Ab, Cry1Ac, Cry1Da, Cry1Cb, Cry7 Aa, Cry8Da, Cry9 Aa, Cry9Ba
israelensis	Cry10Aa, Cry11Aa
japonensis	Cry8Ca, Cry9Da
jegathesan	Cry11Ba, Cry19Aa, Cry24Aa, Cry25Aa
kenyae	Cry2Aa, Cry1Ea, Cry1Ac
kumamotoensis	Cry7Ab, Cry8Aa, Cry8Ba
kurstaki HD-1	Cry1Aa, Cry1Ab, Cry1Ac, Cry1Ia, Cry2Aa, Cry2Ab
kurstaki HD-73	Cry1Ac
kurstaki NRD-12	Cry1Aa, Cry1Ab, Cry1Ac
morrisoni	Cry1Bc, Cry1Fb, Cry1Hb, Cry1Ka, Cry3Aa
tenebrionis	Cry3Aa
tolworthi	Cry3Ba, Cry9Ca
wuhanensis	Cry1Bd, Cry1Ga, Cry1Gb

Apart from sequence similarity, it is apparent that there is a large difference in size between different Cry proteins, though they tend to cluster as either large (~130kDa) or small (~70kDa) proteins. Although there is a considerable difference in size between the different subfamilies, they share a common active core comprising three domains. Simple alignment of different classes of Cry protein, showing the common core domains. It can be seen that the N-terminal end of each gene has a similar organisation, despite the great difference in overall length. In fact, the larger proteins such as the Cry1 group are inactive, and are activated by proteolytic removal of the Cterminal sequence.

The structure of the active core of the Cry1A protein, exemplifying the three discrete domains, each with a quite different folding structure. Domain I at the N-terminal end comprises a series of a-helices arranged in a cylindrical formation. The open cylindrical structure of this domain is thought to be responsible for creating a pore through

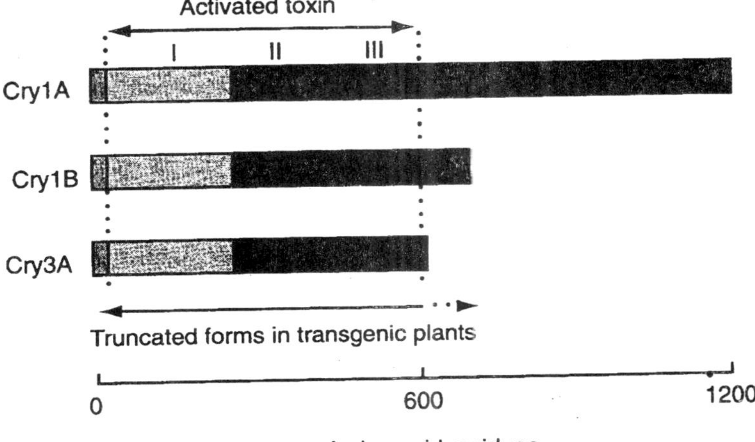

Fig. 8.1. Comparison of the structures of different classes of Cry protein.

the membrane of the insect midgut. Domain II comprises a triple β-sheet and is involved in receptor recognition. Domain III is a β-sandwich, with a number of putative roles including protection from degradation, toxin/bilayer interactions and receptor binding.

The mode of action of δ-endotoxins involves a specific interaction between the protein and the insect larva midgut. After ingestion by an insect larva, the protein crystals are solubilised in its midgut. The larger proteins such as the 130-kDa Cry1 group are proteolytically cleaved at this stage to release the active 55-70-kDa active fragment of the protein. This interacts with high affinity receptors in the midgut brush-border membrane. The result of this binding is to open cation-selective pores in the membrane. The flow of cations into the cells results in osmotic lysis of the midgut epithelium cells, causing their destruction. Thus, the δ-endotoxins are extremely toxic and can be lethal to susceptible insect larvae at relatively low concentrations. On the other hand, their toxicity to other animals (e.g. mammals) is extremely low.

The conditions in the insect larva midgut vary according to insect class. The midgut of Lepidoptera and Diptera is mildly alkaline, whilst the coleopteran gut is generally either more alkaline or acidic. These different conditions favour the solubilisation and activation of different Cry subfamilies. Furthermore, the specificity of the interaction between the endotoxin and the midgut receptor means that individual Cry proteins are active against particular insect larvae.

The Use of *'Bt'* as a Biopesticide

Preparations of *B. thuringiensis* spores or isolated crystals have now been used as an 'organic' pesticide for half a century. The isolated crystals have a limited persistence on foliage of a few days, whilst the spore preparations are effective for about 40 days on foliage and up to 2 years in soil. Neither method of application has a particularly effective penetration, with regard to concealed surfaces and organs of the plant, or against sap-sucking insects—i.e. they are not systemic. However, the agronomic experience of using *B. thuringiensis* spores or isolated crystals as a biopesticide, coupled to the associated safety data and regulatory approvals, has been useful for the rapid development of the' *Bt'* strategy to genetic modification.

Note that *Bt* is generally used as a shorthand for a crop transformed with a *cry* gene (hence *Bt* cotton, etc.), and also for the Cry proteins (hence *Bt* protein). This can be confusing, bearing in mind that we have now met four different terms for effectively the same group of proteins—ICP, δ-endotoxin, Cry and now *Bt*.

Bt-based Genetic Modification of Plants

Although the GM approach to using the *cry* genes to obtain pest resistance in plants is conceptually simple, it does provide an object lesson in the detailed molecular biology that may be required to achieve high levels of expression of a bacterial gene in a transgenic plant. This goes beyond the obvious requirements for plant promoter and terminator sequences to regulate transcription as described in previous chapters. The first attempts to express Cry1A and Cry3A proteins under the control of the CaMV 35S or *Agrobacterium* T-DNA promoters resulted in very low levels of expression in tobacco, tomato and potato plants. It was realised that the prokaryotic gene sequence itself would need to be extensively modified in order to obtain high levels of stable expression. The eventual result was that expression was enhanced by 100-fold to give much better levels of expression, of the order of 100 ng *Bt* protein per mg total protein. Subsequent laboratory tests showed that the effect of producing a specific Cry protein at this level was to provide a considerable degree of protection against damage by susceptible insect larvae. Further confirmation of the effectiveness of these plants in providing protection against insect damage in small scale field trials resulted in the first *Bt* crops gaining approval for commercial planting in the USA in the mid-1990s.

The success of the *Bt* approach led to the development of *Bt* crops by several of the major biotechnology companies involved in crop protection. This table demonstrates the point that the specificity of Cry proteins permits the targeting of specific pests by particular transgenes, and that different crops may have different *cry* genes inserted. We will see later that the nature of the *cry* gene construct can be important for the success or failure of a particular transgenic crop. Indeed, maize and cotton are the only *Bt* crops that are currently commercially grown in the USA, since the New-Leaf *Bt* potato has now been discontinued.

The Problem of Insect Resistance to *Bt*

One of the problems encountered by the accelerating use of *Bt* technology to provide insect resistance has been the equally rapid appearance of resistant pests. This problem initially attracted widespread attention during the first commercial growing season of the *Bt* cotton crop.

The mechanism of insect resistance can be related directly to the specific binding involved in the mechanism of action of the Cry proteins. It may only require a small number of significant mutations in the insect gene coding for the receptor protein to greatly reduce the binding of a particular Cry protein. This is one of the consequences of single gene:gene interactions that will be discussed again in the subsequent chapters on disease resistance. Thus, insect pests resistant to the *Bt* crop could, in theory, appear within a few generations, and repeated growing of *Bt* crops in the same area would provide the selective advantage to accelerate the appearance of a resistant pest population. There are a number of strategies for countering the build-up of insect resistance.

One way is to think of additional GM approaches to tackle the problem caused by the first GM approach. For example, it is possible to use more than one transgene—a process called 'pyramiding', in which transgenes are successively 'stacked' by conventional crosses between different transgenic lines. In order to avoid cross-resistance to two different *Bt* genes, it may ultimately prove better to pyramid *Bt* genes with unrelated resistance genes, such as proteinase inhibitor genes, since it is considered highly unlikely that resistance to two completely different genes will arise in the same organism at the same time.

Another GM approach is to further enhance the effectiveness and range of activities of *cry* genes by, for example, domain engineering to produce chimeric proteins. However, it is still important to realise that *Bt* and other GM approaches to insect resistance cannot be viewed as a 'magic bullet' to permanently eliminate the threat of insect damage. Rather, the GM crops just alter the balance of all the interactions between plant and environment that occur in the field, and the build-up of resistance in the insect population must be managed by good agricultural practices.

The current practice of 'integrated pest management' (IPM) takes into consideration the wider context of a pest/crop interaction (natural predators, adjacent plant species, crop rotation and chemical sprays). For example, selective usage of *Bt* crops (rotating *Bt* crops with *non-Bt* crops, avoiding growing different *Bt* crops within the migration distance of pests common to both, etc.) may prevent the build-up of resistance in the insect population, which would otherwise occur where there is a continuous selective advantage created by repeated use of *Bt* crops. One example of pest resistance management favoured for

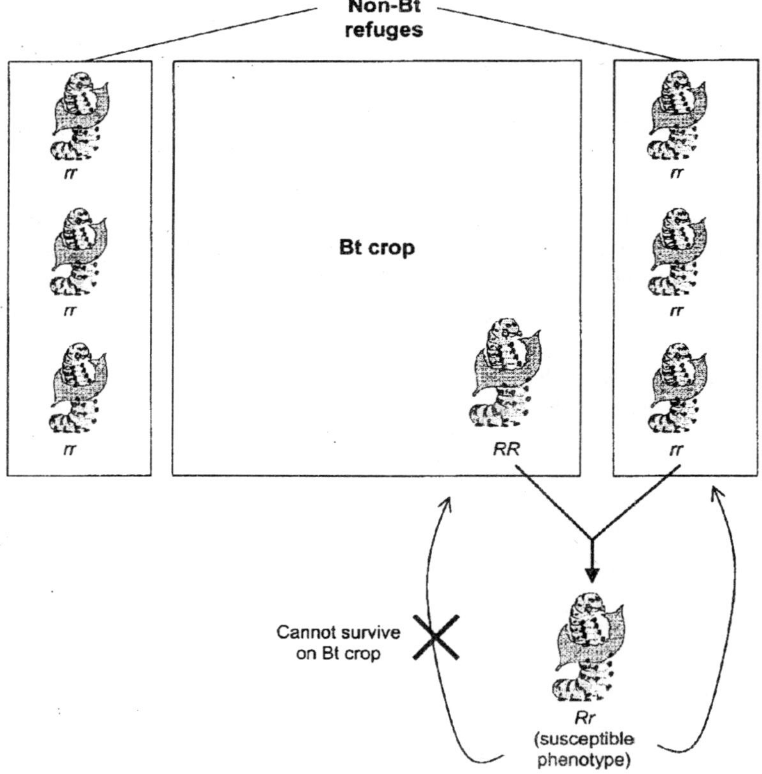

Fig. 8.2. High dose/refuge-resistance management scheme.

managing the build-up of *Bt* resistance is the 'high dose/refuge' approach. In this scheme, transgenic crops expressing a high dose of *Bt* protein are grown alongside a smaller 'refuge' of nontransgenic crop, or any other plant that the insect pest feeds upon. The high dose is important to ensure that only homozygous resistant insects would be able to tolerate feeding on the GM crop. Homozygous resistant insects are likely to be rare in the early stages of the development of resistance. The refuge ensures the presence of a much larger population of susceptible insects in the vicinity of the *Bt* crop than the homozygous resistant insects. Therefore, most mating events between insects carrying one or two resistance genes will probably be with a homozygous susceptible insect, producing heterozygous resistant offspring, which cannot survive on the crop.

It is obviously critical that the approved refuge strategy is correct and monitored, and that the growers are fully informed and comply with the requirements of the resistance management plan. Any deviation from the planting scheme could create the conditions for insect resistance to become established. Another problem resulting from the refuge strategy is that it may provide additional opportunities for the mixing of transgenic and non-transgenic materials.

Environmental Impact of *Bt* Crops

Whilst the first environmental concerns regarding *Bt* crops related to the build-up of resistance in the pest population, a separate issue was brought to the fore by a report that appeared in 1999 indicating that pollen from *Bt* maize might be toxic to the larvae of the Monarch butterfly. The affair does still beg the question of how event Bt176 was approved for commercial growing, given the deliberately high levels of expression of the *Bt* protein in the pollen of these plants. This highlights the need to assess the risks and monitor the environmental impact of GM crops more thoroughly than perhaps has been the case. However, this point also applies to the positive effects of these crops. For example, to what extent have non-pest insect populations benefited from the reduction in chemical pesticides that have certainly resulted from the adoption of *Bt* cotton?

'Copy Nature' Strategy

Some of the problems encountered with *Bt* have led some scientists to contrast the single, highly toxic, heterologous gene approach epitomised by *Bt* with what has been termed a 'Copy Nature' approach. The strategy involves a rational approach to the development of pest-resistant crops and is characterised by the following steps:

Identification of Leads

The point here is to look in nature for plants that show resistance to insect pests. These leads can be found in the literature, world seed collections or observation in the field. The identification of resistant plants provides the starting point for the discovery of plant genes that could confer resistance to insect damage.

Protein Purification

The next stage of the process is the purification of proteins with insecticidal properties. Proteins are the primary target for this type of screen because they provide the most direct way to isolate corresponding genes. (It should be remembered that plants produce a large number of insecticidal secondary products, but it would be a much longer process to engineer the synthesis of a secondary metabolite in a transgenic plant than of an insecticidal protein. Starting from the discovery of a novel compound, the biosynthetic pathway would need to be determined, the key enzymes identified and the genes isolated in order to produce a transgenic plant.) Partial sequencing of the purified protein is one method for isolating the corresponding gene. Characterisation of the protein may also provide a means of identifying the gene in a model organism such as *Arabidopsis*, and using the known DNA sequence as a probe to screen for the orthologous gene in the plant of interest.

Artificial-diet Bioassay

It is important to determine the activity of the isolated protein against the target insect pests by performing feeding assays in the laboratory. In other words, is the protein going to be an effective deterrent to insect feeding at achievable levels of expression in a transgenic plant?

Mammalian Toxicity Testing

Prior to any insertion of the gene into a crop plant, the toxicity of the protein against mammals (and hence potentially humans) should be tested. The point is that it would be a waste of time inserting a gene into plants that subsequently was found to cause concerns about food safety.

Genetic Engineering

It is only at this stage that the isolated gene would be considered valuable for transfer to crop plants. One particular consideration for pest-resistant constructs is the choice of promoter. Although it is possible to rely on a strong, constitutive promoter such as the CaMV

35S promoter, there are certain advantages to using either a tissue-specific promoter or a wound inducible promoter. This is also a 'Copy Nature' strategy, since many insect defence genes are produced site-specifically or are induced by insect damage. Expression of the insecticidal protein at the site of insect damage limits the exposure only to the insects causing the damage. It also reduces the biosynthetic burden on the plant, and reduces the amount of transgenic protein in those parts of the plant destined for human consumption.

Selection and Testing

After transformation, the selection of transgenic plants, confirmation of transformation, inheritance of transgenes in T1, T2, etc. generations and testing of expression levels need to be carried out. The effectiveness of the construct then has to be evaluated by insect feeding assays.

Biosafety

The effect of the transgene on crop yield, insect damage and the wider ecosystem should be properly evaluated in field trials, rather than after several years of commercial planting, as occurred with *Bt* crops.

The aim of the 'Copy Nature' strategy has been stated as: 'insect pest control which is relatively sustainable and environmentally friendly'. The strategy recognises a complex interplay in biological communities between plants, animals, microbes, the soil and the physical environment. It is not just a case of crop plant vs. insect pest. The strategy also takes account of the fact that hostplant resistance to pests is universal and falls into two major categories:

Horizontal resistance

This type of trait is usually polygenic, i.e. it results from the cumulative effect of several minor gene characters. This form of resistance does not involve gene-gene matching (as is the case with Cry proteins and their receptors in the insect midgut, for example) and is usually durable.

Vertical resistance

In contrast, vertical resistance typically arises from one major gene with a high level of expression. The mechanism of resistance may well involve gene-gene matching. The important point to note is that this form of resistance normally occurs in plants to provide a buffer during short-term changes in the level and type of insect damage, and is unlikely to be durable.

CpTI is not the only gene that has been isolated from plants and used to engineer insect pest resistance. Plant resistance genes that

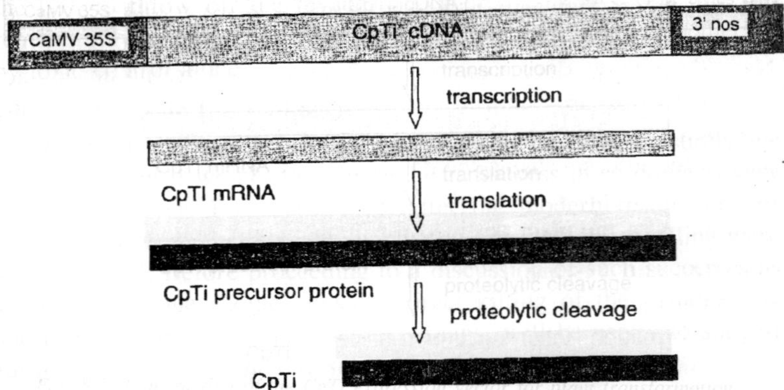

Fig. 8.3. Construction of the CpTI expression vector for plant transformation.

have been tested include other protease inhibitors, α-amylase inhibitors and lectins. However, very few of these examples have progressed from the laboratory to the field trial and none are currently in commercial production. The 'Copy Nature' approach has considerable promise as a complementary method to current pest-management strategies. However, one of the factors which has slowed down the implementation of this strategy is that of food safety.

Insect Resistant Crops and Food Safety

The wide range and type of genes raise understandable questions about their safety for human consumption, given that certain protease inhibitors and lectins are known to have toxic effects in mammals. One of the genes in the table, the snowdrop lectin GNA, was the first

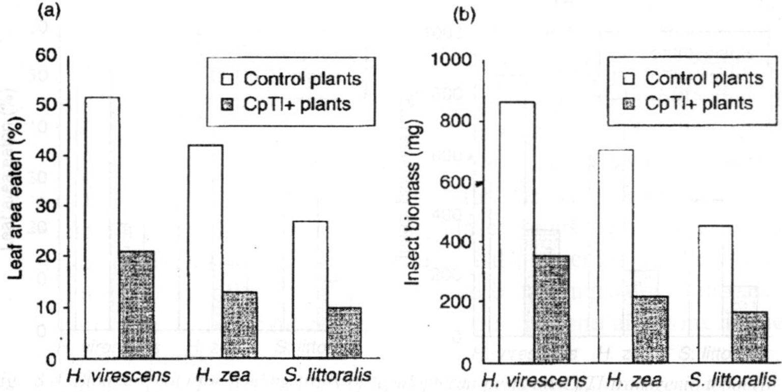

Fig. 8.4. Bioassay data for feeding trails of lepidopteran pests on CpTI transgenic tobacco plants.

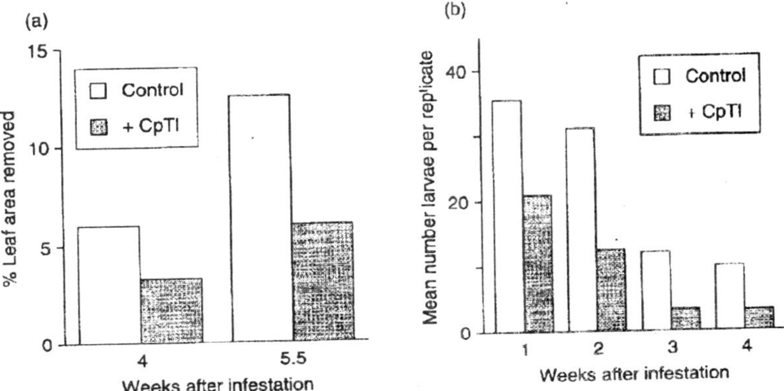

Fig. 8.5. The insect resistance of transgenic tobacco carrying the CpTI gene in the field. The ability of transgenic tobacco to resist damage from Helicoverpa zea larva in the field was assessed by two different assays: (a) the leaf area removed was determined; and (b) the number of larvae per plant was counted.

such gene to attract considerable attention following the suggestion that potatoes carrying this transgene might be responsible for changes to the gut lining of rats.

9

Chemical Control

Weeds have a significant effect on the yield and quality of crops, as a result of competition for light and nutrients, contamination of the harvested crop and because weed populations harbour pests and diseases. Thus weeds are one of the three classes of biotic stress that have a major impact on the proportion of world crop yield available for human consumption. The reduction in world crops caused by weeds, pests and diseases amounts to over one-third of the potential yield, with a roughly equal contribution made by each. Modern agriculture has developed a range of effective herbicides (weedkillers) to tackle the effects of weeds on crop yield. Some of the most useful of these are broad-spectrum herbicides because they are active against a wide range of weeds. However, these can only be used at times when the crop is not itself vulnerable to herbicide action. The development of herbicide-resistant crops therefore offers the opportunity to spray crops at the most effective time to kill weed species without damaging the crop plants. However, it should be noted that some crop plants are naturally resistant to

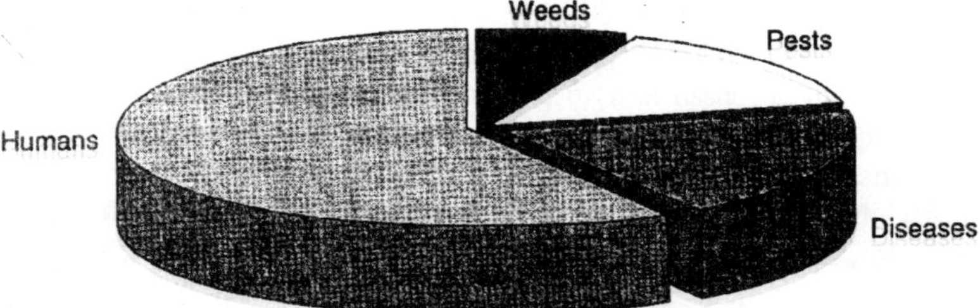

Fig. 9.1. The relative proportions of total world cro, production lost to weeds pests and diseases.

certain herbicides, and that resistant strains may appear through the normal processes of mutation and natural selection. Thus, the concept of herbicide resistant crops is neither novel nor unique to GM technology.

What Types of Compounds are Herbicides?

By definition, herbicides are much more toxic to plants than to animals. Therefore, it is not surprising that they generally affect 'plant-specific' biological processes. Consider for a moment the type of biochemical pathways that are likely to occur in plants but not in animals. Although one's first thought is likely to be of photosynthesis, many other processes are potential targets. Remember that plants are autotrophic, i.e. they can synthesise all their macromolecular components *de novo*. For example, the compounds that are essential in the human diet (vitamins, essential amino acids, etc.) are synthesised by plants, and these biosynthetic pathways are therefore possible targets for herbicides. Many of these pathways are located in the chloroplast, and this has implications for the targeting of many of the transgenic proteins designed to enhance herbicide resistance.

The herbicidal activity of many herbicides has been found to result from the specific inhibition of a single enzyme/protein. It can be seen that herbicides belong to a wide range of different chemical families, with about 15 broad classes of mode of activity. It is also worth noting that most herbicides only have one mode of action, and that certain enzymes seem to be relatively vulnerable to herbicide activity. For example, five chemical families of herbicide target the enzyme acetolactate synthase (ALS). ALS catalyses the first reaction in the branched-chain amino acid biosynthetic pathways, and two common classes of herbicide with widely differing chemical structures (sulphonylureas and imidazolinones) that target this enzyme. As will be seen, knowledge of the herbicide's site of action is essential for some strategies aimed at engineering resistance, but less important for others.

The wide range of chemical families means that herbicides differ greatly in other properties besides their mode of action. For example, they can also be classified according to their:

- site of uptake into the plant (root vs. shoot);
- degree of translocation within the plant (systemic vs. contact);
- time of application (preplanting, pre-emergence, post-emergence or preharvesting).

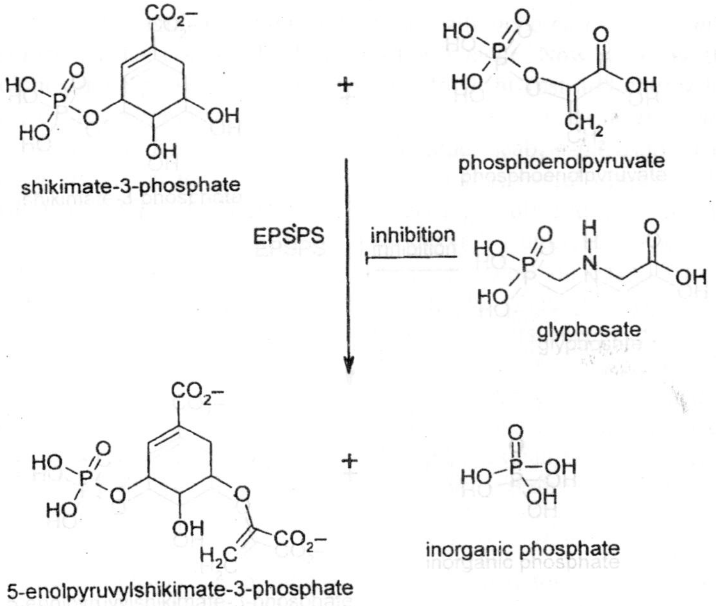
CO2−
HO
HO
OH
OH
shikimate-3-phosphate
+
HO
HO
OH
CH2
phosphoenolpyruvate
EPSPS
inhibition
HO
HO
H
N
OH
glyphosate
CO2−
HO
HO
OH
H2C
CO2−
5-enolpyruvylshikimate-3-phosphate
+
HO−P−OH
OH
inorganic phosphate

Fig. 9.2. Glyphosate and the reaction catalysed by EPSPS.

The widely varied chemical properties of these compounds also means they will differ greatly with regard to toxicity, environmental persistence and biodegradability. The environmental impact of herbicide-resistant crops will be considered at the end of this chapter. However, this is a useful place to make the point that an assessment of the risks and benefits of growing a particular herbicide-tolerant crop must take account of the properties of that particular herbicide.

This again emphasises the point that herbicides are a heterogeneous group of compounds with differing modes of action. In consequence, most transgenic strategies for resistance to a particular herbicide have to be specifically designed for that class of herbicide. Rather than describe in detail all the herbicides against which resistance has been engineered, this chapter will adopt a case-study approach in order to highlight general approaches and strategies.

Strategies for Engineering Herbicide Resistance

The engineering of herbicide resistance demonstrates the way in which quite. different strategies can be used to achieve the same objective. Mullineaux (1992) identifies four distinct strategies for engineering herbicide resistance. These are:

Overexpression of the Target Protein

This strategy effectively involves titrating the herbicide out by overproduction of the target protein. For example, if the herbicide is a specific inhibitor of one particular enzyme, production of sufficient excess enzyme will partially overcome the inhibition. Overexpression can be achieved by the integration of multiple copies of the gene and/or the use of a strong promoter plus translational enhancer to drive expression of the gene.

Mutation of the Target Protein

The logic behind this approach is to find a modified target protein that substitutes functionally for the native protein and which is resistant to inhibition by the herbicide, and to incorporate the resistant target protein gene into the plant genome. Several sources of resistant proteins can be exploited. Note that both the overexpression and mutated target protein strategies require knowledge of the mode of action of the herbicide.

Detoxification of the Herbicide, Using a Single Gene from a Foreign Source

Detoxification is a means of converting the herbicide to a less toxic form and/or removing it from the system. This strategy can be contrasted with the previous two because it does not require a detailed knowledge of the site of action.

Enhanced Plant Detoxification

The aim here is to improve the natural plant defences against toxic compounds. This requires detailed information about endogenous plant detoxification pathways and the mechanisms by which compounds are recognised and targeted for detoxification by the plant.

Prospects for Plant Detoxification Systems

One other strategy that has yet to be fully exploited is the possibility of enhancing endogenous plant detoxification mechanisms. Many xenobiotic (foreign) compounds are detoxified in plants but the pathways may involve more than one step, such as hydroxylation, conjugation and transport stages, so it may prove difficult to identify single-gene mechanisms to engineer resistance. The hydroxylation of compounds involves enzymes such as the cytochrome P450 monooxygenases, which form a large gene family. For example, the analysis of weeds resistant to the herbicide bromoxynil revealed that the bromoxynil was being detoxified by an endogenous cytochrome P450 monooxygenase. This offers the opportunity to use endogenous plant genes to enhance

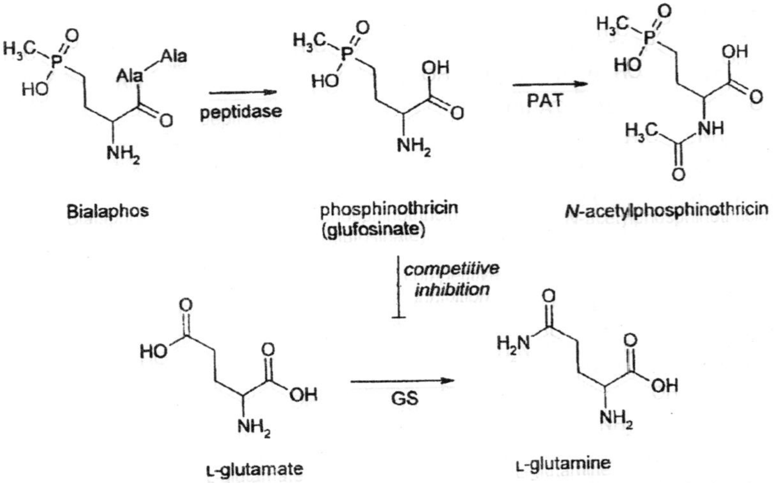

Bialaphos —peptidase→ phosphinothricin (glufosinate) —PAT→ *N*-acetylphosphinothricin

competitive inhibition

L-glutamate —GS→ L-glutamine

Fig. 9.3. The formation, mode of action and detoxification of phosphinothrichin involves removal of the two alanine residues by a peptidase.

resistance against a range of herbicides. However, more research is required to identify which members of the cytochrome P450 gene family are specific for particular classes of xenobiotics, and which have roles in normal metabolic pathways.

Plant detoxification pathways often involve conjugation to glutathione by glutathione S-transferase (GST) activity, and specific transport of the conjugate into the vacuole. GSTs also comprise a large gene family, some members of which are known to be involved in endogenous metabolic reactions. In some cases, the hydroxylation and conjugation pathways operate in concert. Hence, the resistance of maize to atrazine is ascribed to a two-step pathway involving both 2-hydroxylation and conjugation to glutathione. Thus, there is the potential here to enhance endogenous systems, or transfer systems between plant species, once more information about the functions of these large gene families is available. The exploitation of functional genomics techniques will accelerate the acquisition of this knowledge.

Commercialisation of Herbicide-resistant Plants to Date

This chapter has described the two most widespread examples of genetically manipulated herbicide resistance. It is important to note that some of these have not been widely planted commercially. The atrazine-resistant crops, for example, have not been developed further, given the environmental concerns about the use of this class of persistent herbicides.

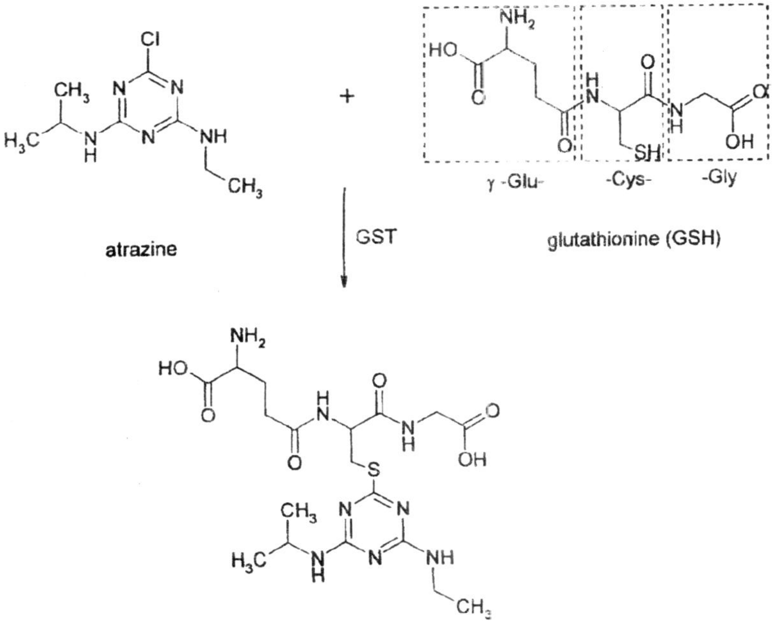

Fig. 9.4. The conjugation of atrazine to glutathione. The detoxification of atrazine is a two-stage process involving a 2-hydroxylation step prior to the addition of glutathione.

All these examples use either the mutant target enzyme approach, or the prokaryotic detoxification gene approach, which have been described in detail when considering the glyphosate and phosphinothricin case studies. However, one more example is worth exploring further, because it demonstrates the use of homologous recombination to modify an endogenous plant gene to engineer herbicide resistance.

The Environmental Impact of Herbicide-resistant Crops

The predominance of herbicide-resistant crops in terms of their development and commercial growing was highlighted at the beginning of this chapter. Some of the reasons for this trait maintaining its position in the GM crops 'league tables' have been discussed. Scientifically, herbicide resistance has a number of advantages that facilitated its rapid development. A number of different single-gene strategies using accessible genes are possible, and the trait itself can be used as a selectable marker. These factors, combined with the research and development impetus provided by the agrochemical industry, inevitably led to their prime position. However, it should be noted that herbicide-resistant crops do not top the league table just because they had such a favourable 'head start'. They have maintained their leading position because of the rapid adoption of these crops in countries where the technology has been accepted.

In the USA, where the greatest area of GM crops are grown, the proportion of herbicide-resistant soybean rose from 17% of the soybean acreage in 1997 to 68% in 2001. Similarly, herbicide-resistant cotton expanded from 10% of the cotton acreage in 1997 to 56% in 2001. This rapid adoption of herbicideresistant crops begs a number of questions. The first one is whether farmers are adopting these crops so rapidly because there is a clear commercial advantage for them, and if so, what is it? The second one relates to the environmental impact of this rapid switch in growing patterns, and hence herbicide usage patterns. This indicates that although the use of glyphosate increased compared to that for non-transgenic soybean, the total use of herbicides decreased. Indeed, the major beneficiary of this technology was the farmer rather than the producer, mainly by way of a marked overall reduction in herbicide costs. Thus, the assumption that herbicide-resistant crops will inevitably lead to a greater use of herbicides needs to be qualified. The use of the target herbicide may well increase, but this may displace or reduce the requirement for other herbicides, such that there is a net reduction in their use. Since the environmental impact of herbicides such as glyphosate and phosphinothricin is thought to be lower than many of the compounds they are displacing, the producers of herbicideresistant crops can claim that the adoption of their crops will have a positive effect on the environment. However, there are other environmental concerns about the rapid adoption of herbicide-resistant crops. One area of concern is the reduction of biodiversity as a result of the more efficient removal of weed species from arable land. These concerns relate more widely to current intensive agricultural practices, and lie outside the direct scope of this book. Another area of concern relates to the possibility of encouraging the de velopment of herbicide-resistant weeds, and is discussed below.

The Development of 'Super Weeds'

One of the major concerns about the use of herbicide-resistant crops is that their introduction will lead to the appearance of 'super weeds', which could evade control by the commonly used herbicides. It is important to put these fears into context. We have previously discussed the fact that some crops are already resistant to certain herbicides, and that herbicide-resistant weeds may appear naturally wherever there is a selective pressure created by the repeated use of the same herbicide. Thus, the problem of herbicide-resistant weeds is not a new phenomenon unique to GM technology and, bearing in mind the large armoury of herbicides is not one that need pose a major threat. Nevertheless, it is sensible to ask whether the widespread use of herbicide-resistant crops will make a difference to the number, rate of appearance and type of resistant weeds because: (1) they encourage the repeated use of the same herbicide, and (2) the herbicide resistance gene is transferred to a weed population by one of the processes of 'gene flow' described below.

Herbicide-resistant weeds could theoretically arise by three types of mechanism:

1. The herbicide-resistant crop itself appears as a 'volunteer' weed in fields where rotational crops are grown. This volunteer population could reproduce outside of cultivation and form a self-sustaining weed population.
2. Pollen from the herbicide-resistant crop fertilises weedy the crop plant, producing herbicide-resistant hybrids.relatives of Subsequent backcrossing of these hybrids with the weed species could lead to introgression of the herbicide-resistant trait into the weed population.
3. There is horizontal gene transfer (as opposed to the vertical transmission in points 1 and 2) by other mechanisms (e.g. viruses) that spreads the herbicide-resistant trait into a much wider range of plant species.

Just as the environmental impact of each herbicide should be considered separately, it is also important to treat different crops on a case-by-case basis. Thus, certain crops are normally more prone to form volunteer weeds than others, and the presence of a herbicide-resistant strain could be a problem if that herbicide is normally used to clear the volunteers.

However, the large arsenal of herbicides should ensure that volunteer weeds can be cleared from cultivated land for rotational crops. It must be borne in mind that "weediness" is a complex mix of characteristics that do not normally include herbicide resistance. When considering gene flow from a GM crop to other crops and weed species, the nature of each individual crop must also be considered. For example, the spread of pollen from out-crossing crops is more of a problem than with crops that are self-crossing. Another factor to consider is the proximity of closely related weed species. Thus, maize, soybean and cotton have no compatible wild-type relatives in the countries where GM varieties are widely grown (USA and Canada). Wheat, oilseed rape and sugar beet are grown in close proximity to related species, but only the latter two are out-crossing. For this reason. oilseed rape and sugar beet have been used as "worst case analysis' crops for the study of gene flow into weeds. The studies that have been carried out to date tend to indicate that there is detectable gene flow from crops to weed species, but that transgene introgression under field conditions is probably rare. In order to minimise this potential gene flow, it is necessary to predict the distance that pollen can travel from each GM crop, and to establish appropriate buffer distances to prevent the fertilisation of weedy relatives. However, many of the studies to determine buffer distances have been on a relatively small scale. A recent large-scale study using herbicide-resistant oilseed rape (produced by non-transgenic means) does indicate that oilseed rape pollen can travel much further than had been suspected, but that nevertheless, the amount of gene flow to nonresistant plants was minimal.

There is as yet little evidence that horizontal gene flow between plants poses a particular problem for the transfer of transgenes into unrelated species. What is clear about the entire issue of environmental impact of herbicide resistant crops is that much more research is required to establish the full extent of all types of gene flow between plants in the environment. Thus, a recent review of the environmental risks and benefits of GM crops concludes that:

1. The risks and benefits of GM crops are not entirely certain or universal.
2. The ability to predict ecological impacts of any introduced species (GM or not) is difficult, and available data have limitiations.
3. Some benifits and risks may exist that have not yet been identified or addressed in pulblished literature.
4. The quantity and quality of different GM crops that may eventually be developed merit special consideration for risk assessment.
5. Better evaluation of potential benifits will help risk managers know how these balance with potential risks.
6. Measures developed to prevent gene transfer to wild plants can reduce the potential environmental impacts and prolong potential benifits.

10

DISEASE CAUSING ORGANISM

There are several important plant diseases, which are thought to be due to mollecutes. Some of the most common, widespread and economically important diseases are described in this chapter. Diseases caused by non-helicle, non-cultivated MLO include: Alfalfa witches' broom, Banana bunchy top, Clover phyllody, Sesamum phyllody, Sunflower phyllody, Coconut lethal yellowing, Eggplant little leaf,Legume little leaf, Elm yellows or phloem necrosis. Groundnut witches' broom,little peach, Mulberry dwarf, papaya bunchy top,peach decline, X diseases of peach, Potato purple top roll, Potato marginal flavescence, Potato witches' broom, Sandal spike, Sugarcane grassy shoot, Sugarcane white leaf, Tomato big Bud, Tomato purple top, Yellow leaf of areca palm and Yellow wilt of sugarbeet. The diseases are characterized by yellowing, chlorosis, or bronzing of foliage, stunting (shortening of internodes and reduced size of leaves), proliferation of axillary buds resulting in witches' brrom effect, proliferation of secondary roots and abnormal fruits and seeds. Flowers are often sterile. Some diseases show few of these symptoms whereas others may show all or most of these symptoms. Variation in symptom syndrome may occur according to host species, the time and mode of infection, environmental factors, and mixed infection by more than one type of phloem-limited pathogens such as dual infection of MLO and Spiroplasma or MLO and RLB or other bacteria or even virus. Established spiroplasma diseases also show most of these symptoms but virescence is not caused by spiroplasmas. The disease caused by *Spiroplasma* spp. Include Bermuda grass white leaf, Citrus stubborn (*S. citri*), corn stunt (*S. kunkelii*), Opuntia witches' broom, Rice yellow dwarf, and yellows of aster and lettuce.

Aster Yellows Disease

Aster yellows is a disease that affects over 300 species of plants, including flowers, vegetables, and weeds. Susceptible flowers include aster, chrysanthemum, cockscomb, coreopsis, cosmos, daisy, dianthus, echinacea (coneflower), gladiolus, marigold, petunia, and phlox. Susceptible vegetables include carrots, onions, potatoes, and tomatoes. Weeds such as dandelions, plantain, and thistle are also susceptible and can serve as a source of inoculum in home gardens.

Causal organism: Phytoplasma

Vector

Leafhoppers. The six-spotted leafhopper is the main vector, which transmits the *Phytoplasma* that causes aster yellows in flax and in many other crops such as canola and sunflower, and in some weeds.

Disease cycle

Aster yellows is primarily transmitted by insects called leafhoppers, and through grafting. The phytoplasma survives winter in perennial and biennial plants. In the spring, leafhoppers feed on infected plants. During feeding, plant sap containing the phytoplasma is sucked into the leafhopper's body where multiplication of the microorganism takes place. Following an incubation period, the phytoplasma is transferred to healthy plants during leafhopper feeding. When a leafhopper feeds on a plant infected with aster yellows it becomes "infected" with the phytoplasma and remains infected throughout its life. The phytoplasma cells multiply and cause infection of the insect's salivary glands within one to three weeks. When the infected insects feed on healthy plants, they inject the phytoplasma cells into the plant phloem. Susceptible plants will be symptomatic in 10 to 40 days.

The spread of aster yellows is worse in cool wet summers. Hot dry weather is not favorable for either the phytoplasma or the leafhopper. As with many disease and pest problems, diagnosis is perhaps the most important factor in controlling aster yellows. The cause of aster yellows is a microscopic organism called a phytoplasma (formerly called a mycoplasma-like organism or MLO). Phytoplasmas are most similar in size and composition to bacteria.

Symptoms

Specific symptoms vary from plant to plant. Symptoms of aster yellows are more severe and appear more quickly during warm weather. At lower temperatures, plants may be infected without symptom expression. There are following Aster yellows symptoms:

1. The roots are bitter, stunted, and deformed with tiny, hair-like roots developing all over the main root.
2. Infected plants may also produce a proliferation of spindly, secondary shoots, giving the plant a bushy appearance.
3. Yellowing of the top part of the plant, yellowing of the foliage, Carrots and marigolds can be severely affected by aster yellows. Leaves of infected carrots grow in tight bunches. The inner leaves are yellow and stunted, while outer leaves turn rusty red to reddish purple. Infected marigolds have small, yellow leaves.
4. Chlorosis, yellowing of the leaves while the veins remain green, is a major symptom of aster yellows. Growth slows down and leaves may be smaller and more narrow than usual. Foliage is sometimes curled.
5. The symptoms of the disease will often differ depending upon what species is infected. For instance, carrot roots may be bitter and hairy while lettuce may show pink or tan spots and have twisted inner leaves.
6. Production of many spindly stems and/or flower stalks.
7. Conspicuous malformation of the flowers, all flower parts including the petals is converted into small, yellowish green leaves.
8. Diseased flowers are sterile and produce no seed. Flowers often fail to develop color, remain green and distorted, and seeds or fruit never develop. Flowers that are produced are often irregular, deformed, and green. Flowers may be deformed and exhibit bizarre tufts of deformed leaves inside the flower or in place of the flower.
9. The severity of the disease depends on the stage at which plants become infected and the number of insect vectors that carry the organism. The whole plant is generally stunted and spindly.

Control measures

1. Once a plant is infected with aster yellows, it is a lost cause since the disease is incurable. Early diagnosis and prompt removal of infected plants may help reduce the spread of the disease. Although the disease itself is not fatal to the plant, its presence makes it impossible for a plant to fulfill its intended role in the garden.
2. Controlling aster yellows is difficult. As long as infected leafhoppers are around, they can infect plants. A practical way to avoid having problems with this disease is to grow plants that are not as susceptible to aster yellows. Verbena, salvia, nicotiana,

geranium, cockscomb, and impatiens are among the least susceptible plants.

3. Vegetable growers may protect susceptible crops by using the mesh fabrics that keep leafhoppers and other insects away from the plants. Some growers put strips of aluminum foil between rows because bright reflections of sunlight confuse the leafhoppers.
4. Remove weeds in lawn, garden, and surrounding areas, including plantain and dandelion that may harbor the disease. Insecticides are generally not recommended for the control of leafhoppers in the home garden.
5. Seed early to avoid the migrating leafhoppers in mid and late season. Seeding early also reduces the incidence and severity of aster yellows.

Papaya Bunchy Top Disease

Papaya bunchy top has been observed throughout much of the Caribbean region, from Cuba southward to Trinidad, since it was first reported in Puerto Rico in 1931. Bunchy top is a devastating disease and can severely limit commercial papaya production.

Causal organism: *Phytoplasma* (MLO)

Vector

Leafhoppers. Two leafhoppers, *Empoasca papayae* Oman and *E. stevensi* Young, transmit the bunchy top agent, and the occurrence of the disease coincides largely with that of these vectors. *Empoasca papayae* has most frequently been associated with the disease. *Empoasca papayae* has rarely been found on other plant species and is the only leafhopper known to breed on papaya. *Empoasca stevensi* was originally described from specimens collected from papaya in Florida in 1940, but little is known about the ecology and distribution of this species.

Disease cycle

Single insects of both vector species can transmit the bunchy top MLO. The MLO also can be transmitted by grafting. Symptoms usually begin to appear 30 to 45 days after inoculation. Some papaya cultivars are more tolerant to the disease than others, but immunity is not known. The degree of tolerance affects the rate and severity of symptom expression. However, it is not known whether vector preference and the ability to transmit the bunchy top MLO are affected by cultivar differences. Although bunchy top was originally thought to be caused by a virus, the disease is most likely caused by a mycoplasma-like organism (MLO). MLOs have been observed in the phloem of infected plants using transmission electron microscopy, and infected plants treated with tetracycline-based antibiotics exhibited a remission of symptoms. Although the absence of latex flow from fresh puncture wounds in leaves, petioles, stems, and fruits were considered diagnostic for bunchy top, a more recent study indicated that this technique is unreliable. The presence of MLOs in plants with foliar symptoms of bunchy top and normal latex flow was confirmed by electron microscopy. The DNA-binding fluorochrome, 4,6-diamidino-2-phenylindole (DAPI), can be used to stain MLOs in the phloem of numerous plant species, thus, making their presence detectable by epifluorescent microscopy. This technique has been used to detect MLO's in petiole and leaf vein tissue of papaya trees with bunchy top symptoms. The reliability of this method to diagnose bunchy top has not been established.

Symptoms

1. The first symptom of bunchy top is a faint mottling of the upper leaves.
2. The laminae of infected leaves progressively become more chlorotic, especially in interveinal areas, and may eventually exhibit marginal necrosis.
3. Leaves and petioles show reduced growth, and become rigid.
4. Internodes become progressively shortened, and petioles tend to assume a more horizontal position.
5. Oily appearing spots are often present in the upper parts of stems and in petioles.
6. Apical growth ultimately ceases, which, with the shortening of internodes, imparts a "bunchy top" appearance to affected plants.
7. With the cessation of apical growth and decline of plant vigor, the oldest leaves progressively abscise, leaving fewer, more stunted leaves at the top. Eventually, plants may be entirely killed; however, new shoots may sprout from the lower, non-affected portion of stems.
8. If fruits are set on infected plants, their flavor may be bitter due to the disease.

Control measures

1. Currently, the use of tolerant cultivars is the only practical means of control of bunchy top, but it has limited application. Where bunchy top occurs, tolerant cultivars can only be grown commercially in locations with low disease pressure. Disease

pressure varies with geographic location, presumably due to insect vector ecology.

2. Application of insecticide to reduce the incidence of leafhopper vectors may be beneficial.
3. In the past, the disease has been managed through a program involving leafhopper control throughout the productive life of the crop and the removal of sources of inoculum. Rouging infected trees or the topping of infected plants below the point where latex exudes after wounding attained the latter. Axillary shoots that develop after topping are often free of the disease. However, the success of this program depended largely on the availability of a pesticide with long residual activity, namely DDT, which is no longer approved for use in many areas. A suitable replacement for DDT has not been identified.
4. Antibiotic therapy is an effective control measure from an experimental standpoint but has not been applied commercially.
5. Drenching the soil around infected plants with chlortetracycline or tetracycline hydrochloride and drenching combined with root-dip treatments were both successful. The efficacy of foliar application or trunk injection has not been established.

X-Disease in Peaches

X-disease has been diagnosed in several peach orchards in worldwide. Once the disease is established in an area or orchard, it can be very destructive.

Causal organism: X-disease mycoplasma

Vector: Leafhoppers.

Disease cycle

Peach X-disease often follows a 10-15 year cycle. The disease will be very severe for 4-6 years, then will gradually disappear as a commercial problem until the next cycle begins. A good diagnostic symptom is the presence of apparently healthy twigs or branches with normal-looking leaves mixed with twigs or branches showing the symptoms described earlier. This mixture of healthy and diseased branches on the same tree occurs primarily during the first and second years of infection. Two to three years after initial infection, most branches will show symptoms. X-disease is caused by a phloem-limited mycoplasma - a minute pathogenic organism smaller than most bacteria. The X-disease mycoplasma is transmitted by at least eight species of leafhoppers. White apple leafhopper, rose leafhopper, and potato leafhoppers are not vectors. In fact, none of the X-disease vectors are abundant enough to cause direct feeding damage and they usually escape notice in the orchard. However, they are very efficient vectors of X-disease. The leafhoppers acquire the X-disease organism while feeding on diseased chokecherry bushes, on infected sweet cherry trees, or on wild seedlings of sweet cherry. They do not acquire the X-disease mycoplasma from diseased peaches because the population of the disease organism within diseased peach trees is so low that leafhoppers do not encounter the organisms while feeding. X-disease can be transmitted by several species of leafhoppers. These leafhoppers acquire the X-disease pathogen while sucking juices from the leaves of X-disease-infected choke-cherries. Wild chokecherry is an important reservoir for the X-disease mycoplasma. Two to three weeks later the leafhopper can inject the agent into healthy leaves while feeding. These leafhoppers are usually not considered pests of peach and cherry. The main damage they cause is the incidental transmission of X-disease. Movement of leafhoppers, and therefore X-disease, can also occur from infected sweet and sour cherry, particularly from trees. Although the possibility of spread from peach to peach has been investigated many times, it appears to be of minor importance.

After leafhopper vectors feed on an infected plant, the X-disease organism must grow within the insect for at least 20 days before the insect can transmit X-disease to another plant. Once that 25-day incubation period is completed, however, the leafhoppers with X-disease remain infective for the rest of their lives. In laboratory studies, leafhoppers have often lived for 30-40 days after they become infective. A single infective insect therefore has the potential to infect numerous plants. Peach trees that become infected with X-disease are usually inoculated by leafhoppers during July and August, and symptoms then develop on the trees the following year. Leafhoppers can continue to transmit X-disease to peach trees during September and early October, but many of the late-season transmissions fail to cause disease because the pathogen does not become established in the plant following late-season transmissions. Relatively mild winters during the last two years may have allowed more of the late-season transmissions to persist through winter and may therefore have contributed to the increasing severity of X-disease.

Symptoms

Symptoms on peach may be variable, depending upon the environment, variety, and possibly other factors. However, once acquainted with general symptoms, one should be able to recognize the disease more often than not.

1. Symptoms are predominantly foliar, but the fruits may also be affected. The first symptoms occur on foliage. X-disease causes leaves on infected peach trees to turn yellow, curl upward, and develop red, water-soaked spots that are not limited by leaf veins. The leaf disorder results in early defoliation of the oldest leaves, leaving a "horse-tail" of young foliage at the end of affected terminal shoots.
2. Leaves on affected branches fall prematurely, starting at the base of the branch. Eventually, only a tuft of leaves remains at the tips of infected shoots.
3. Fruit set on trees infected with X-disease may appear normal at first, but fruit on infected branches will usually drop prematurely.

Control measures

1. Reducing populations of leafhopper vectors is an approach to X-disease control. Maintaining a vigorous insect control program from June through harvest with insecticides effective against leafhoppers should be useful in reducing spread of the disease.
2. Remove infected trees near peach orchards.

Leaf Variegation or June Yellows Disease in Strawberries

Although leaf variegation or June yellows in strawberries was noted and described as "gold-striped" and "silver-striped" as early as 1719, the cause of the disorder is still not known. Various synonyms for the disorder are Blakemore yellows, spring yellows, chlorosis, and noninfectious variegation. Leaf variegation occurs commonly in, but is not limited to, strawberry cultivars that have Blakemore, Howard 17 (Premier) or Auchincruive Climax in their parentage.

Causal organism: Mycoplasma-like organism (MLO)

Vector: Grafting or sap inoculation.

Disease cycle

No causal agent has been associated with leaf variegation or June yellows. The disorder has not been transmitted asexually by grafting or sap inoculation but can be transmitted sexually to seedlings of affected parents. Its appearance is unpredictable in all breeding lines and may not become evident in a particular clone for several years. The current theory is that leaf variegation is due to some entity within the strawberry host cells that acts similarly to certain virus or mycoplasma infections.

Electron microscopy has thus far not identified a virus or mycoplasma in diseased plants.

Symptoms

1. Symptoms first appear in the spring on new unfolding leaflets. A puckering and distortion may occur in leaves with a "white streak" type of variegation in early stages of the disease.
2. The leaves become irregularly mottled, streaked, or spotted with golden or pale yellow-to-white and light green areas. A slight loss of plant vigor accompanies the color change.
3. Variegation occurs mainly during cool weather, when the temperatures are below 50°F (10°C), in the spring or fall, although some cultivars may remain mottled throughout the summer.
4. Variegation has been observed only on the foliage and flowers-never on fruit. All runners and daughter plants produced by diseased plants are also variegated with no reversion to a normal green color.
5. Seemingly healthy green plants may become variegated at any stage. There is no way to predict when a plant or its offspring will become variegated.
6. As the disease progresses, the symptoms increase in intensity. The leaves become progressively more mottled each year until they are completely golden yellow-to-white and frequently puckered or otherwise distorted.
7. White-streaked or fully yellowed leaves never regain a normal green color during the summer.
8. Affected plants become dwarfed.
9. Fruit production is reduced considerably, with the fruit being small and of poor quality.
10. Affected plants never recover and usually die within two to three years.
11. Permanent white streaks or sectors, with or without accompanying mottling symptoms, may occur in some cultivars affected by leaf variegation or June yellows. White streak may be a chimera (the result of a somatic mutation). It occurs more commonly in some cultivars, for example, in Earliglow, than in others.

Control measures

1. Do not propagate from variegated mother plants.
2. The only control to save a variety from extinction is to locate a source of "yellows-resistant" or normal plants of the desired variety. These plants, if found, should be propagated by virus-free procedures

to increase the number of healthy plants. Never accept plants that show leaf variegation.

3. The disorder appears to be noninfectious in that normal green clones may grow beside variegated clones for long periods without becoming variegated.
4. Where the percentage of variegated plants in new or established plantings is reasonably low, variegated plants should be rogued (removed and destroyed) as soon as the disease is detected and replaced with nonvariegated, certified, virus-free plants.
5. At present there is no cure for leaf variegation, and there is no assurance that healthy green plants in a variegation-susceptible cultivar will remain nonvariegated.

Corn Stunt

Causal organism: Mycoplasma, *Spiroplasma kunkelii*

Vector: Corn leafhopper (*Dalbulus maidis*)

The adult corn stunt leafhopper and nymphs are light colored. Adults fly or jump when disturbed and often gather on the corn plant in the center of the whorl. Nymphs have no wings and readily move when disturbed. Early planted corn is rarely affected. Leafhoppers suck juices out of leaves with their stylet mouthparts. High populations can cause premature leaf drying, and black sooty mold may grow on the honeydew. The presence of leafhoppers however does not necessarily mean corn stunt occur. The insect only develops high populations in some years and vectors the corn stunt pathogen even less often.

Disease cycle

Corn stunt disease severity is vary depending on the stage of plant development when infection occurs. Corn stunt is caused by a mycoplasma, *Spiroplasma kunkelii*. Mycoplasmas differ phenotypically from other bacteria by their minute size and lack of a cell wall. The disease is transmitted to a plant when the vector, the corn leafhopper (*Dalbulus maidis*), feeds from the plant phloem. The corn stunt mycoplasma is introduced into a healthy plant with the insect saliva. In the plant, the mycoplasma is restricted to the phloem, but spreads throughout the plant infecting all areas. Corn leafhoppers carry the disease from infected corn to healthy corn. *S. kunkelii* is not transmitted via plant seeds, nor are the leafhopper eggs infected. The leafhoppers feed on an infected plant for several days before acquiring the spiroplasma. It then requires a two to three-week incubation period before the leafhopper can infect a healthy plant. It cannot survive outside its host and is dependent upon transmission for survival and dispersal.

Symptoms

1. The infected plants are stunted.
2. Leaves have broad chlorotic strips and with age they take on a reddish-purple colour.
3. Internodes are reduced in length.
4. The infected plants bear numerous small ear shoots that usually are poorly filled.

Control measures

Resistance to corn stunt has been identified and is available in some hybrids. These hybrids have shown a 0.5 to 0.75 tons/acre grain yield increase over a susceptible hybrid.

Tomato yellows Disease

In some years, serious outbreaks of tomato yellows have occurred in southern British Columbia. The severity varies from year to year; with a high infection 30%, or more, of the crop may be lost. The disease is widespread throughout the Pacific coast region extending from British Columbia to Mexico.

Causal organism: Beet curly-top mycoplasma.

Vector: Beet leafhopper.

Symptoms

1. Diseased plants are readily distinguishable in the field, because they are rigid and have stiff brittle leaves that tend to roll inward along the midrib.
2. They curve downward slightly and have a drooping but not wilting appearance. The veins are generally, but not invariably, purplish and the normally green leaves change to sulfur yellow.
3. The infected plant ceases to grow and the fruit already formed does not enlarge.
4. The pith dries out and the stem becomes hollow. Eventually the plant turns brown and dies.
5. The disease runs its course in 1-2 weeks. Affected plants may recover somewhat but never regain a normal appearance. The small roots become brown and decayed.

Control measures

This disease is caused by the same mycoplasma that causes curly-top of beets; therefore always separate crops of tomatoes from beets. Maintain good weed control.

Purple-top Disease of Tomato

Causal organism: Aster yellows mycoplasma

Symptoms

1. The most characteristic symptom is the purplish tinge in the foliage, which is particularly apparent in the youngest leaves.
2. Affected plants show an upright appearance because of slight hardening of the stem and leafstalks.
3. The leaves also show a tendency to roll slightly.
4. Considerable stunting of growth occurs.

Control measures

Purple-top disease is rarely serious enough to warrant specific control measures. Good weed control is always sound husbandry; it removes sources of inoculum.

Yellows Diseases

Plant pathogenic microorganisms called phytoplasmas are the major cause of yellows diseases in plants. "Yellows" diseases are the most important group of diseases spread by phytoplasmas. Phytoplasma causing the yellows diseases of echinacea, monarda and caraway described.

Causal organism: Phytoplasmas

Vector: Phloem-feeding leafhoppers.

Disease cycle

Phytoplasmas are transmitted between plants primarily by phloem-feeding leafhoppers (Insecta: Homoptera: Cicadellidae). These insects have a well-developed sensory system that enables them to reject unsuitable host plants without sampling the sap. They usually transmit phytoplasmas most efficiently to plants preferred for feeding. Leafhoppers feed on a variety of plants and can spread phytoplasma pathogens from wild to cultivated plants and vice versa. Infected biennial and perennial plants may allow phytoplasmas to persist from one year to another in fields and natural areas.

Leafhoppers can survive through the winter as eggs and appear as early as May. The incidence of phytoplasma diseases often depends upon the abundance and time of migration of leafhoppers from infested to healthy plants. Traditional field crops, such as alfalfa and potato, may serve as symptomless carriers of phytoplasma pathogens and also harbour adult leafhoppers, which could transmit primary inoculum to adjacent plantings of specialty crops.

Symptoms

1. Affected plants generally show leaf yellowing, reddening and stunting in the early stages of disease development.
2. The yellowing symptom can be quite striking, and the name "yellows" is often used to describe these diseased plants.
3. Plants with yellows sometimes have extremely numerous, small, branched, axillary shoots coming from the stem nodes, giving them a bunchy or witches'-broom appearance. In later stages of symptom development, infected flowers may change into vegetative, leaflike structures (virescence and phyllody) and, consequently, fail to produce seeds.
4. Yellows can also reduce winter hardiness, and it often kills affected plants.

Control measures

1. An integrated strategy, focussing on both the pathogen and the vector, must be followed to successfully manage phytoplasma diseases in susceptible crops. Susceptible specialty crops should not be planted close to fields of forage legumes or other perennial crops affected by phytoplasma diseases.
2. Care should also be taken to control perennial weeds that may harbour phytoplasmas and/or leafhoppers both in susceptible crops and in adjacent fields and around their margins.
3. Specialty crops and surrounding fields and headlands should be regularly monitored for leafhoppers.
4. Keeping leafhopper populations low is an important means of controlling phytoplasma diseases.
5. Registered insecticides can be used to control leafhoppers in some situations.
6. Development of resistant cultivars is the most promising method for controlling phytoplasma diseases in plants, but little effort has been devoted to breeding specialty crops with resistance to phytoplasmas.

Potato Purple-top wilt MLO Disease

This disease is particularly Occurs in North America and, in so far as it causes potato haywire disease, is considered as an A1 quarantine pest by EPPO (OEPP/EPPO, 1984). All but the first of these are regarded as belonging to "the aster yellows" group, with yellow or purple leaf discoloration as main symptoms. They are not usually transmissible by tuber. The leafhopper vectors include Macrosteles and Hyalesthes spp. In Europe, there are numerous MLOs of the aster yellows type, attacking various hosts, but only potato stolbur MLO is found in potato. It is usually considered as a distinct MLO. Tomato big bud MLO occurring in Australia, several Asian countries and USA causes a disease very similar to stolbur. The three Indian MLOs are also recognized as of the Old World aster yellows type and are distinguished from potato stolbur MLO. This disease, reported for the first time from India, can cause 15 to 80 percent loss in yield.

Causal organism

Mycoplasma Like Organism (MLO) of the aster yellows complex.

Vector

Leafhopper Macrosteles fascifrons and nymphs of Alebroides nigroscutellatus (*A. dravidanus*)

Disease cycle

Under natural conditions, potato purple-top wilt MLO can be dispersed by its leafhopper vectors. Internationally, it could be carried in potato plants or in vector insects associated with them, but such movement hardly occurs in practice. It should be stressed that aster yellows type MLOs are not normally tuber-transmitted in potato. The principal vector is the leafhopper Macrosteles fascifrons, in which the MLOs are propagative. The vectors remain infective for life. They can feed on a variety of plants, and the apparent host range of the MLO depends more on the host preferences of the vector than on any particular specificity between plant and MLO. Potato purple-top wilt MLO from Japan is reported to be transmitted by Scleroracus flavopictus and not by Macrosteles spp., unlike aster yellows MLO from USA. Potato MLOs vary in their size and shape. They are pleomorphic with a diameter up to 1000 nm. They occur mainly in the sieve tubes and can be seen there by electron microscopy or fluorescent detection methods. In India, Nymphs of Alebroides nigro-scutellatus (A. dravidanus) are better transmitters of the pathogen. The MLO adversely affects its vectors. The urge for mating in the females is delayed and often they become sterile. The hatching time of eggs is enhanced. The longevity of adults is also reduced. The minimum acquisition access period is one hour. Maximum transmission takes place with acquisition access period of three dyas. The MLO needs 7 to 15 days of incubation period in the vector body. One or a few infective vectors are required per plant for maximum transmission. Nymphs are more efficient than adults in transmitting the MLO. While adult males have better transmitting ability than females, the latter retain the MLO in their body for a longer time. Infected tubers seem to be the most important source of perennation and transmission. The symptoms are pronounced at low temperatures of 150 to 200^{0}C and masked at 300 to 380^{0}C. The MLO can be transmitted to clover, tomato, Datura and tobacco.

Symptoms

1. Apical leaves are pinched or curled, roll inward and develop yellowish-purple pigments.
2. Infected plants lose their apical dominance and proliferation of axillary buds occurs.
3. Plants may wilt and die prematurely. These plants commonly form hair sprouts.
4. According to Conners (1967) the primary symptom is purple-top wilt, and the haywire stage is due to secondary infection.
5. It is characterized by rolling and purple or pink colouration of the basal part of the leaflets of top leaves. Some cultivars develop only rolling but no colouration. Stunting, chlorosis, profuse axillary shoots with aerial tubers, and swelling of nodes are other symptoms. There is no wilting of the infected plants and no necrosis of the tubers.

Control measures

The diseases are not important enough to merit any particular control measures.

Grassy Shoot Disease of Sugarcane

Grassy shoot, yellowing and white leaf (albinism) diseases of sugarcane in India, Taiwan and Thailand have been considered as the same disease caused by the same organism. Grassy shoot was first noticed in India in 1919 in Maharashtra and is one of the serious diseases of sugarcane there. It has been found in Andhra Pradesh, Tamil Nadu, Orissa, Bihar, Uttar Pradesh, Uttaranchal, Punjab and Rajasthan. Outside India, Grassy shoot is reported from Burma, Sri Lanka and Sudan also.

Causal organism: Mycoplasma Like Organism (MLO)

Vector

Rhopalosiphum maidis (*Aphis maidis*), *Aphis sacchari*, and *Aphis indosacchari* and leafhopper.

Disease cycle

The organism is present in the sieve tubes of the phloem as ovoid, spherical, or irregularly shaped bodies. The size of ovoid or spherical bodies is 300 to 400 nm in diameter and that of protruding filaments 30 to 50 nm in diameter. These cells are devoid of true cell wall and are bounded by a single triple layered membrane. They contain ribosomal granules and DNA strand.

Mostly the cells are concentrated towards the periphery of the cell, near the cell-walls but often the cells are fully loaded with these bodies and then the cell dies followed by death of the MLO due to starvation. In some instances MLO are seen lying close to sieve pores of the sieve tubes and turn filamentous while passing through the sieve pores. Concentration of MLO is high in canes growing at high temperature (around 300^{0}C) and such canes show severe symptoms. At lower temperatures the symptoms are less severe and the number of MLO in the cells is also lower.

The white leaf disease of sugarcane reported from Taiwan has some difference from grassy shoot disease described in India. While grassy shoot organism is inactivated by hot water at 500^{0}C for two to three hours the white leaf disease agent is inactivated by hot water at 550^{0}C for 30 minutes or 540^{0}C for 40 minutes. Aphids transmit the grassy shoot while leaf hoper transmits white leaf.

The grassy shoot MLO perpetuates through diseased canes used for seed and spread through diseased sets and cane cutting knives (sap transmission). Ratooning of a diseased crop is an important source of perennation. Many aphids and a leafhopper have been reported to spread the MLO from plant to plant in the field but only to a limited extent. The vectors are *Rhopalosiphum maidis* (*Aphis maidis*), *Aphis sacchari* and *Aphis indosacchari*. The minimum acquisition feed period is 34 seconds and inoculation feed period 30 seconds. However, optimum acquisition feed period is 15 minutes. Although a single infected aphid can transmit the pathogen, optimum number is 15. Transmission is in the non-persistent manner. Winged forms are more effective in spreading the disease in the standing crop. The disease has been transmitted from sugarcane to sorghum and from sorghum to sugarcane by using Aphis sacchari (*Melanaphis sacchari*). Sorghum is also a natural host of the MLO.

Symptoms

The disease shows different types of symptoms at different stages of plant growth and, thus, has been variously described as 'new chlorotic disease', 'albino disease', yellowing disease', 'bunchy disease' or 'leaf tuft', Profuse tillering is the main symptom and, therefore, the standard name for the disease is grassy shoot.

1. The most pronounced symptom is the grassy appearance of the affected stools.
2. Shoots growing from diseased setts remain dwarfed or stunted.
3. The leaves are narrow and small, like grass leaves; the canes are thin with short internodes, giving a bunchy or grassy appearance to the clump.
4. The leaves appear yellowish and in some cases may be entirely devoid of any pigment (white leaf or albinism). If many of the tillers are affected in this manner the entire stool dries.
5. In secondary infection, ordinarily the new leaves and internodes show the above symptoms. Leaves exhibit straight, long, white or light green or yellowish streaks.
6. The lower nodes produce large number of grassy shoots. In systemically infected canes the disease appears in May-June. In sprouts raised from top buds the symptoms appear late as compared to sprouts raised from lower buds. This suggests difference in the concentration of the causal agent in the cane.

Control measures

1. Healthy seed sets should be taken from a field where grassy shoot is not present even in traces.
2. During the early stage of crop growth insect vectors can be controlled by weekly spraying of 0.16 per cent malathion.
3. Diseased clumps should be dug out and destroyed.
4. Heat therapy has also been recommended. The sets should be dept in hot water at 500°C for two hours or in hot air at 540°C for eight hours. This treatment is feasible only for seed nurseries.
5. Tetracycline antibiotics (250 ppm) applied to seed sets under negative pressure have been found to eliminate symptoms. Single bud sets can be treated with such antibiotics.

Little Leaf of Eggplant (Brinjal)

Little leaf of brinjal (*Solanum melongena*) is found throughout India and Sri Lanka and probably occurs in other neighbouring contries. In India it was first reported from Coimbatore (Tamil Nadu). The disease has become a serious threat to the profitable cultivation of this vegetable crop in most of the states. When young plants are attacked they do not produce flowers and fruits.

Causal organism: Mycoplasma Like Organism (MLO)

Vector: *Cestius phycitis* or *Hishimonus (Cestius) phycitis*

Disease cycle

In nature, the causal agent of brinjal little leaf is transmitted by the vector *Cestius phycitis* or *Hishimonas phycitis*. The same MLO occurs on *Datura fastuosa* and *Vinca rosea*. Artificially it has been transmitted to tomato, potato and tobacco. Probably, during the off-season of the brinjal crop the causal agent survives on weed hosts and from there it is transmitted to the main host by its insect vector.

Symptoms

1. The main symptom of the disease is production of very short leaves by the affected plant.
2. The petioles are so much reduced in size .that the leaves appear sticking to the stem.
3. Such leaves are narrow, soft, smooth and yellowish in colour. Newly formed leaves are further reduced in size.
4. The internodes of the stem are shortened and at the same time a large number of axillary buds are stimulated to grow into short branches with small leaves. This gives the whole plant a bushy appearance.
5. Usually such plants fail to form flowers. Even if flowers are formed they remain green. Fruiting is rare.

Control measures

No effective control measures against the disease are known.

1. Atibiotic treatments with tetracyclines have been claimed to control the disease.
2. In the absence of any definite control measure eradication of weed hosts and diseases brinjal plants and control of insect vectors by insecticides are the only recommendations against the disease.

Sandal Spike Disease

Sandal (*Santalum album* L.) is an angiospermic semi-root parasite of many plant species, which grows to the size of a tree. It is one of the most economically important forest trees in India. The species is confined to India and Indonesia. In India the tree is common in the southern parts including states of Karnataka, Tamil Nadu and Kerala. Spike disease is the most destructive of the few diseases that attack this plant. It is a yellows type of disease with witches' broom effect. Trees of all ages and size are attacked and die within a few years of appearance of sysmptoms. The disease is most common in Karnataka but has also spread to parts of Tamil Nadu and Kerala.

Causal organism: Mycoplasma Like Organism (MLO)

Vector

Dodder, root contacts, insect vectors and leafhoppers *Nephotettix virescence* and *Moonia albimaculata*.

Disease cycle

The spike disease is transmitted through root contacts and insect vectors. More than one species of leafhoppers seem to be involved in its transmission. These include Nephotettix virescence and Moonia albimaculata. The size of MLO in the host ranges from 60 to 750 nm, the most commonly occurring form being ellipsoidal, measuring 180-220 × 250-300 nm. Some cells are elongated, measuring up to 750X150nm. The MLO grows best at 300 to 3800°C. Below and above this range the growth declines. The MLO is reported to infect a large number of plant species such as *Eucalyptus grandis*, *Vinca rosea*, *Zizyphus oenoplea*, *Dodonea viscose*, etc. The disease can be transmitted from vinca rosea to sandal and vice versa through dodder.

Symptoms

Generally two types of symptoms are produced. These are 'rosette spike' and 'pendulous spike'.

1. The more common symptom is 'rosette spike', in which found severe reduction in leaf size and shortening of the internodes. This results in crowding of leaves on the leaf bearing branches.
2. The new leaves that develop later are further reduced in size. The reduction is more in width than in length. Such leaves stand out stiffly on the branches like spikes.
3. In advanced stages of the disease, just before death of the tree, leaves become yellowish and finally reddish.
4. The flowers of the infected trees show phyllody (virescence).
5. The diseased parts rarely bear any fruit although the healthy parts of the same tree bear normal flowers and fruits.
6. Ends of roots of diseased plants die and haustorial connections with the host plant are damaged.
7. Phloem necrosis is one of the internal symptoms.
8. The 'pendulous spike' is another characterized symptom of this disease in which continuous apical growth of individual shoots without proportionate thickening of the shoot resulting in a drooping habit of the shoot.
9. The dormant buds do not develop or grow and, therefore, no resetting effect is seen. In this form of the diseased roots and haustoria are not damaged.

Control measures

So far no specific measures for control of sandal spike disease have been developed.

1. Attempts had been made to cure the trees by raising temperatures above 380^{0}C or by boiling fires in trenches around the affected trees but it was neither successful nor practical.
2. Eradication of diseased trees and creating disease free strips around affected areas have also failed to arrest its spread. No host resistance has been found.
3. Like other plant diseases of MLO etiology sandal spike also shows remission of symptoms after treatment with tetracycline. The systemic fungicide Benlate was also reported to bring about temporary remission of symptoms.
4. Application of the antibiotic by the girdling method shows recovery of the trees 25 to 30 days after treatment with terramycin alone or in combination with Benlate. However, these treatments result in only temporary relief for 50 to 150 days and then the disease reappears.

Citrus Greening Disease

Citrus greening disease is a major cause of crop and tree loss in many parts of Asia and Africa. Before it was identified as one disease, it became known by various names: yellow shoot (huanglungbin) in China; likubin (decline) in Taiwan; dieback in India; leaf mottle in the Philippines; vein phloem degeneration in Indonesia; and yellow branch, blotchy-mottle, or greening in South Africa. As it became clear that all these were similar diseases the name "greening" was widely adopted. In India, this is one of the most important and widely prevalent mycoplasmal diseases of Citrus. Indian dieback was first accurately described in 1929 and attributed to poor drainage.

Causal organism: Mycoplasma Like Organism (MLO)

Vector

Graft Transmissible and psyllid vectos, *Trioza erytreae, Diaphorina citri*.

Disease cycle

Greening has been experimentally transmitted by dodder to periwinkle (*Catharanthus roseus*) in which it induced marked yellowing. As in its citrus hosts, African greening in periwinkle required temperatures below 27°C for symptom development, whereas the Asian form was more heat-tolerant. The dodder itself is a host for the greening pathogen with titers higher than has been observed in citrus. Research on Asian greening followed close behind. In 1966 Salibe and Cortez demonstrated graft transmission in the Philippines, and reported on an insect vector, soon identified there and in India as another psylla, *Diaphorina citri*. The number of adult psylla of either species in a population that carry the disease is relatively small, but under experimental conditions, a single adult of either species can transmit greening. Other psylla species have been noted on citrus, namely *T. eastopi* (*T. litseae*) and Mesohomatoma lutheri in Réunion, D. communis in India. Organisms were observed in the hemolymph and salivary glands of *T. erytreae*, the phloem of yellow shoot and likubin-infected citrus and in infective *D. citri*.

Symptoms

Symptoms can occur throughout the tree, especially if the infection occurs at or soon after propagation. If infection occurs later the symptoms and the causal organism are often partially confined.

1. Infected trees or branches suffer heavy leaf drop followed by out-of-season flushing and blossoming, with dieback occurring in severe cases.
2. In general leaf symptoms are of two types: Primary symptoms are characterized by yellowing of normal-sized leaves along the veins and sometimes by the development of a blotchy-mottle. With secondary symptoms the leaves are small, upright, and show a variety of chlorotic patterns resembling those induced by zinc and iron deficiencies. Analysis of symptomatic leaves shows higher potassium content and lower calcium, magnesium, and zinc concentrations.
3. Infected fruit are small, lopsided and have a bitter taste, probably because of higher acidity and lower sugars. Many fall prematurely, while those that remain on the tree do not color properly, remaining green on the shaded side, hence the name of the disease. Any seeds in severely affected fruit are often abortive.
4. Root systems are poorly developed with relatively few fibrous roots, possibly because of root starvation. New root growth is suppressed and the roots often start decaying from the rootlets.
5. Amino acid concentrations are generally lower except for L-proline, which is higher in the leaves and flower bud. There are also increases in amylase activity and glucose and fructose production, increased respiration early in infection and the appearance of gentisoyl-β-D-glucose, which seems to inhibit peroxidase activity.

These changes do not, however, appear to be involved in symptom production.

6. Cytopathic studies have revealed pockets of necrotic phloem, excessive phloem formation, abnormal cambial activity and accumulation of starch in plasmids. Cytoplasmic membranes formed from in vaginations of the plasmalemma, aberration of the chloroplast thylakoids and mitochondrial collapse have also been observed. These changes indicate significant metabolic disturbances, possibly via toxins and/or hormones.

Control measures

The following control measures can be used for controlling the disease.

Thermotherapy

In 1964 Lin reported eliminating yellow shoot disease by water-saturated hot air treatment of graftwood at 48-58°C with no loss of tissue viability. Other attempts to eliminate Asian greening with heat have had varying degrees of success. In India treatment of budwood at 47°C for 2 h reduced disease incidence and longer treatments eliminated the pathogen. Treatment of infected young plants or seedlings budded with infected tissue at 38-40°C for three to four weeks also killed the pathogen. In South Africa budwood from infected trees heated over a hot water bath at 51°C for 1 h, 49°C for 2 h and 47°C for 4 h eliminated the disease although some tissue viability was lost at the higher temperatures. Infected trees covered for 2-5 months with polyethylene-covered fiberglass sheets showed a dramatic decrease in the number of diseased fruit. However this method is impractical for large-scale use. Heat treatments have some application in the elimination of greening from horticulturally desirable trees, possibly in conjunction with shoot tip grafting.

Chemotherapy

The association of prokaryotic organisms with greening prompted investigations into the use of antibiotics. Martinez et al. in the Philippines suppressed leaf symptoms with a foliar spray of tetracycline hydrochloride and apparently eliminated greening in budwood by immersion in the solution. In India immersion in a penicillin-carbendazin dip reportedly gave complete control. In South Africa Schwarz and van Vuuren injected various tetracyclines into the trunks of greening-infected sweet orange trees and found that tetracycline hydrochloride gave the best results, reducing fruit symptoms in the next crop from 60 to 20%. The incidence of infection remained unchanged for three years when a severe psylla outbreak and a slight increase in greening occurred. The trees received a second treatment that reduced fruit symptom incidence to below 10%. Better results were achieved when two injections were given one month apart, while a 97% reduction of fruit symptoms was achieved when trees were injected continuously under pressure for 7 days. The best time for injection is spring and distribution in the tree can be enhanced by the use of hyaluronidase. Trunk injections of tetracycline hydrochloride have also been successful in Taiwan, China, Réunion and the Philippines. Other antibiotics have been tested. Penicillin had a less suppressive effect than tetracycline hydrochloride in Asia and Réunion, but was totally ineffective in South Africa. In India partial to total remission occurred following injection with a compound called BP-101 (Hindustan Antibiotics Ltd.), ledermycin (demeclocycline hydrochloride) and a streptomycin-chlortetracycline mixture, while complete remission using foliar sprays of agrimycin and carbendazin has also been reported.

However, tetracycline hydrochloride is phytotoxic. Buds immersed in concentrations of over 250 ppm did not survive after grafting. Van Vuuren observed leaf narrowing, lamina yellowing, veinal browning, and occasional defoliation of individual twigs. A brown discoloration was found in the trunk in a zone extending some 300 mm from the hole, decreasing in intensity with distance. Leaf narrowing has also been recorded in Taiwan with oxytetracycline, but not with tetracycline hydrochloride or chlortetracycline.

A derivative of tetracycline hydrochloride, N-pyrrolidinomethyl tetracycline (PMT) is more soluble in water, giving slightly better control of greening and causing no foliar phytotoxicity and minimal wood discoloration. Midsummer treatments gave the best results. The solubility of PMT in water is 1250 mg/ml, compared to 10.8 mg/ml for tetracycline hydrochloride, so that smaller volumes can be injected from syringes without employing high pressures.

Moll was unable to detect by fluorimetry any antibiotic residues in the juice of injected trees one month after treatment. Other studies using growth inhibition of *Bacillus megaterium*, *B. cereus* var. mycoides or *B. subtilis*, have confirmed that antibiotic residues drop rapidly.

Use of insecticides

Because psylla are sap feeders, systemic insecticides are the most effective. Those most commonly used in southern Africa are endosulfan sprays; monocrotophos applied as sprays, aerial sprays, trunk applications and trunk injections; and dimethoate applied to the soil. Weekly sprays of monocrotophos did not prevent infection, while high volume and mist blower applications destroy natural enemies of other citrus pests. A single application of dimethoate by microject and drip irrigation gave control for over 7 weeks, and two applications gave season-long protection. Simultaneous injection of dimethoate and tetracycline hydrochloride controlled psylla without reducing antibiotic efficacy. In India, *D. citri* has been controlled for up to 4 weeks with sprays of several insecticides, including dimethoate and monocrotophos and soil applications of dimethoate. In China field control with soil applications of O-methoate has been reported. However insecticides can interfere with biological control of other pests. While dimethoate can reduce the efficacy of the biological of citrus red scale, at 0.01% a.i. it does not seriously affect the activity of psylla parasites.

Biological control

The major parasite of *T. erytreae* in South Africa is a species of parasitoid wasp Tetrastichus that oviposits in the psylla nymphs and appears to limit psylla populations. A survey in Réunion in 1973 found no parasitized mummies of psylla and it was suggested that parasitic wasps should be introduced. Three species, Tetrastichus radiatus (now called *Tamarixia radiata*) from India and *Tetrastichus dryi* and *Psyllaephagus pulvinatus* from South Africa, were collected and mass reared. *T. dryi* was released in 1976, and had dramatically reduced the *T. erytreae* population by 1978, with complete eradication by 1982. *P. pulvinatus* failed to establish itself. *T. radiata*, which is specific for *D. citri*, was introduced in 1978 and by 1982 had virtually eliminated psylla from commercial orchards, surviving only on *M. paniculata* hedges. The incidence of greening has consequently been reduced. Both species of wasp were subsequently introduced into Mauritius where they reduced psylla populations, but more slowly than in Réunion, possibly because they were not mass reared before release. *T. radiata* has also been introduced into Taiwan, the Philippines, Nepal and Indonesia, but without much apparent success. Other parasitic wasps of *D. citri* are *Diaphorencyrtus aligarhensis*, *Psyllaephagus* sp., *Chartocerus walkeri*, and *Encarsia*, but none appears to be a potential biological control agent. The efficacy of parasitic wasps is limited by the activity of hyperparasitic wasps. In South Africa the effect of Aphidencyrtus cassatus varies according to its ability to maintain synchrony with its host. Another species is *Cheiloneurus cyanonotus*. In Taiwan *D. aligarhensis* has at least 10 hyperparasites, of which one, Pachyneuron, can also oviposit in *T. radiata*. In China a species of Tetrastichus attacks both *T. radiata* and *D. aligarhensis*. No hyperparasites occur in Réunion, which may in part explain the success of biological control there. Fungi provide another means of biological control. *T. erytreae* is attacked by *Cladosporium oxysporium* and *Capnodium citri* , but both are sensitive to desiccation and are density-dependent. *D. citri* is parasitized by *Beauveria* and *Cephalosporiurn lacanii*, the latter somewhat effective at high density.

Other control measures

The production of greening-free trees is vital. The use of shoot-tip grafted material can ensure this goal. Using sticky yellow traps to identify nursery and orchard sites that are less susceptible to psylla attack can facilitate production. Such sites should also be distant from indigenous psylla hosts. Although psylla were considered not to have strong dispersal powers, recent investigation established that adult *T. erytreae* can be wind-dispersed in the absence of host plants 1.5 krn or more, and can survive approximately 85 hr without food. Strong winds such as those associated with typhoons have transported *D. citri* over medium to long distances. The removal of infected branches or trees and neglected trees may also reduce inoculum sources although mean population densities in the latter are not always significantly higher. If winter irrigation is withheld from lemon trees to suppress new flush, such infection reservoirs can be reduced. In China hedges of *M. paniculata*, which serve as hosts for *D. citri*, are being removed.

Sesamum Phyllody Disease

This disease is quite common in sesamum growing areas. Though the disease has been known since a long time, Pal and Pushkarnath (1935) first reported it as a graft transmitted virus disease.

Causal organism: Mycoplasma Like Organism (MLO)

Vector: Graft Transmissible and *Orosius albicinctus*.

Disease cycle

This disease is transmitted by *Orosius albicinctus*. The incubation period of the pathogen in leafhoppers was found to be 15-63 days and 13-61 days in the host (*Sesamum*) depending upon the atmospheric conditions, especially the temperature. Both male and female leafhoppers were found to be equally infective. The disease could be transmitted through vectors or grafting to a number of plants of field crops, ornamentals and common weeds. Studies on the biology of the vectors showed that they can over winter at all stages and are present in nature throughout the year. The pathogen has a very wide host range.

The important hosts are *Brassica Campestris* L. vars., toria, sarson, *Brassica rapa* and gram.

Symptoms

1. Vein clearing of the leaves, phylloid flowers, stimulation of axillary buds, which results in profuse branching, reduction in leaf size.
2. In affected plants, floral parts are transformed into green, leaflike structures resulting in complete sterility.
3. Shortening of internodes are some of the characteristics symptoms of the disease.

Control measures

The diseases are not important enough to merit any particular control measures. Late sowing of the crop may reduce in the incidence of the disease.

Rice Yellow Dwarf Disease

This disease was first observed in Japan in 1910. Now it is widespread in many South-east Asian countries. In India this disease has been reported from many parts of Orissa, Bihar, West Bengal, Andhra Pradesh and Delhi.

Causal organism: Mycoplasma Like Organism (MLO)

Vector: Green hoppers of rice *Nephotettix* spp.

Disease cycle

The disease is known to be transmitted by green hoppers of rice *Nephotettix* spp. The causal organism of this disease has been found to be associated in the phloem cells of the infected leaf tissues.

Symptoms

1. The disease is characterized by pronounced stunting and tillering with uniformly pale green to pale yellow chlorotic leaves.
2. Early infected plants die prematurely, while the late infected plants may survive but remain sterile. Sometimes they may bear a few poorly developed flowers and grains.

Control measures

Attempts have been made in japan to control the vectors of this disease by aerial application of insecticides over large areas. Seed dressing with systemic insecticides, Carbofuran and Propoxur can protect rice seedlings from the rice green leaf hoppers, the common vectors for both virus and rice yellow dwarf.

Elm Yellows (Phloem Necrosis)

Causal organism: Phytoplasma

Vector: Phloem-feeding insects as Leafhoppers.

Disease cycle

Elm yellows is caused by a phytoplasma (formerly called a mycoplasma-like organism). The disease is believed to be spread by such phloem-feeding insects as leafhoppers. The phytoplasma overwinters in infected tree roots and witches' brooms.

Symptoms

Symptoms of elm yellows may appear any time during the summer but are most common in mid- to late summer.

1. Symptoms of elm yellows include yellowing and drooping of foliage followed by leaf drop and death of branches.
2. This pattern may occur on one or a few branches or may quickly involve the entire tree.
3. Susceptible trees may show symptoms over the entire tree in a matter of a few weeks. Tolerant trees become stunted and may develop bunchy, prolific growth at the tips of branches (called witches' brooms) or on the trunk.
4. The inner bark tissues of infected trees often exhibit a butterscotch or light brown discoloration in small streaks or flecks.

Control measures

The diseases are not important enough to merit any particular control measures.

Witches'-broom Disease of Bean

Witches'-broom is caused by a species of mycoplasma but is of little economic importance in beans. However, it is worth mentioning because it represents a totally different category of disease.

Causal organism: Mycoplasmas

Vector: Leafhoppers and dodder plants.

Disease cycle

Mycoplasmas are harbored in perennial and biennial weeds and in eggs of infected leafhoppers in which they can propagate, passing through the eggs. Leafhoppers are the most important vectors that transmit the disease, but there are other means of disease transmission, namely by grafting scions and by parasitic dodder plants.

Symptoms

1. On beans, the infected plant is dwarfed and the internode shortened.
2. The infected plants produce extra shoots, and thus the plants are shooty and leafy.

3. Petioles are borne at wide angles, and flower peduncles are often elongated and wavy.
4. The leaves may curl slightly and later become somewhat yellow.
5. Pod sets are largely reduced. However, death seldom occurs in the beans.

Control measures

Control of leafhoppers by insecticides and removal of perennial and biennial weeds is the key to avoiding this disease.

Witches' Broom Diseases of Red Clover

Causal organism: Mycoplasma-like organism (MLO)

Vector: Leafhoppers

Disease cycle

The causal organism is transmitted by some kinds of leafhoppers.

Symptoms

1. Mycoplasm-like organism disease causing shrinkage of the plant.
2. The disease mainly occurs in red clover. A lot of small leaves at first appear from the center part of the plant.
3. These leaves turn to reddish brown or fade.
4. The plant gradually begins to rosette and wither.

Control measures

Control of leafhoppers by insecticides and removal of perennial and biennial weeds is the key to avoiding this disease.

Stunt Disease of Bushes

Causal organism: Mycoplasma-like organism (MLO)

Vector: Leafhopper *Scaphytopius magdalensis*

Disease cycle

Stunt is a serious disease caused by a mycoplasma-like organism (MLO). Almost all cultivars are susceptible to stunt. Only Rancocas cultivar has resistance. The sharpnosed leafhopper (*Scaphytopius magdalensis*) spreads this disease.

Symptoms

1. The disease causes a stunting of bushes.
2. Leaf symptoms are the best evidence for disease diagnosis. Leaves show downward cupping and yellowish margins.
3. Normal green coloration of the leaf is usually retained along the midvein and lateral veins. Leaf size is usually reduced.
4. Berries are smaller than normal on infected bushes.

Control measures

Remove and burn infected bushes. The normal full-season insecticidal control program for blueberries usually keeps the leafhopper population in check. Replant with disease-free stock.

Prunus X Disease

This is also known as Eastern peach X disease, Western peach X disease.

Causal organism: X Disease Phytoplasma

Vector: Leafhoppers

Symptoms

1. Symptoms appear six weeks after foliation. Blotches develop across the leaf blades; first appearing water-soaked then turning red/yellow.
2. The rest of the leaf becomes chlorotic with margins rolled upward. Leaves become tattered as the blotches drop out.
3. Affected leaves fall early, leaving only small rosetted leaf tufts at the shoot tips. In early stages, symptoms are limited to few branches; the disease then progresses randomly.
4. Most fruits drop soon after the first symptoms develop. Those remaining are angular and color early. They produce shriveled, nonviable seeds, and discolored flesh around the pits.

Control measures

The diseases are not important enough to merit any particular control measures.

Peach Yellows Disease

This is also known as Peach virus 1, Prunus virus 1, Chlorogenus persicae caused disease.

Causal organism: X Disease Phytoplasma

Vector: Leafhoppers

Symptoms

1. Leaves are chlorotic, yellow, and often roll upward; red spotting develops as leaves begin to droop.
2. Severe infections will develop slender shoots with small, narrow, yellow leaves.
3. Buds that are normally latent on younger trees produce dwarfed leaves.
4. Terminal dieback is common in one-year-old shoots. Trees usually die 2 to 3 years following the appearance of symptoms.

5. Fruits are of poor quality and tend to reach maturity three weeks earlier than normal.
6. Red-colored cultivars have saturated coloration, which may be spotted. The flesh exhibits red streaks or marble patterns, with redness around the pit.

Control measures

The diseases are not important enough to merit any particular control measures.

Peach Vein Clearing Disease

A study was conducted from 1982 to 1990 in the Pyrenees-Orientales to investigate susceptibility of the Rouge du Roussillon apricot cultivar to apricot chlorotic leaf roll phytoplasma diseases (ACLR). The trees in the study were found to have the following atypical symptoms, and have been identified as peach vein clearing.

Causal organism: Unknown phytoplasma.

Symptoms

1. In summer, mild chlorosis and partial rolling of the leaves were apparent while the veins seemed to be juttying.
2. Vegetation is prolonged in autumn.
3. The number of superior sized fruit is reduced. Trees do not decline.
4. In winter, bark became swollen and bud burst was delayed from 6 to 8 days.

Control measures

The diseases are not important enough to merit any particular control measures.

Peach Rosette Disease

This is also known as Peach rosette virus, Peach virus 2, Prunus virus 2, Carpophthora rosettae, Nanus rosettae, Persicavirus rosettans caused diseases.

Causal organism: X disease strain phytoplasma.

Symptoms

1. Initial symptoms develop quickly. The first leaves reach normal size but become yellow and roll inward or arch backward, their veins sometimes appearing red. Defoliation of these leaves follows.
2. Internodes of new shoots are extremely short with leaves pressed into dense rosettes.
3. As leaves on older parts of the trees have already fallen, only tufts of young leaves on the tips of bare shoots remain.

4. Blossoms rarely develop and/or the flowers do not set fruit on affected branches.
5. Most diseased trees die within the first year of infection; those that leaf out the following spring die shortly thereafter.

Control measures

The diseases are not important enough to merit any particular control measures.

Peach Red Suture Disease

This is also known as Peach virus 5, Prunus virus 4, Persicavirus rubescens caused disease.

Causal organism: X disease phytoplasma.

Symptoms

Symptoms caused by the disease are generally attributed to a strain of peach yellows phytoplasma.

1. Fruits of affected trees mature and soften several days earlier than normal on their suture sides, while their converse sides remains green and hard.
2. The suture side often becomes swollen; its flesh, coarse and watery.
3. Fruits may develop irregular contours, and in red-fruited cultivars, the suture side remains a deeper red with purple areas at the apexes.
4. Leaves exhibit a yellowish-green or bronze color, noticeable after petal fall and before harvest.
5. Foliage also develops autumnal coloring earlier than normal.
6. On two-year-old branches, buds may grow into spur-like outgrowths.

Control measures

The diseases are not important enough to merit any particular control measures.

Peach Little Peach Disease

This is also known as Peach virus 3, Prunus virus 1A, Chlorogenus persicae var. micropersica caused disease.

Causal organism: X disease strain phytoplasma.

Symptoms

1. Leaves of infected trees are distinctly greener than normal during the early part of the vegetation period.
2. Proliferation on side branches and spurs of 2 to 3-year-old branches gives the tree a bushy stature.
3. Shoots are long and upright.

4. Leaves are leathery and droop downwards, later becoming chloroitic.
5. Symptoms first appear on one part of the tree, and then spread through its entirety by the third year.
6. Maturation of the small and malformed fruits is delayed, sometimes as much as by 3 weeks.
7. Their pits are small, their kernels underdeveloped.
8. Flavor of affected fruits is insipid.

Control measures

The diseases are not important enough to merit any particular control measures.

Molieres Disease

A rare disease affecting sweet cherry and European plum, Molieres disease appears to be restricted to southwestern France.

Causal organism: European Stone Fruit Yellows Phytoplasma.

Symptoms

1. Initial symptoms appear as slightly chlorotic leaves in summer.
2. The following year the trees bloom abundantly, though flowers are malformed and fruit set is poor.
3. The few fruit that do develop remain small and drop prematurely.
4. Small deformed leaves, rosetting and poor lignification of young shoots as well as phloem and bark necrosis also are associated with the disease.
5. Young shoots die back and as the disease progresses the entire scaffolds die.
6. In new orchards, diseased trees are usually not observed until 3 to 7 years after planting.
7. All cherry cultivars and rootstocks, especially *P. mahaleb*, appear to be susceptible, however, plum cultivars differ in susceptibility.

Control measures

The diseases are not important enough to merit any particular control measures.

Cherry Decline Disease

Causal organism: Unknown phytoplasma.

Symptoms

Symptoms are attributed to a complex of the Prune dwarf ilavirus and Strawberry latent ringspot nepovirus, which induce Cherry Rasp Leaf (European).

1. Enations, in the form of leafy outgrowths or small protuberances along the midrib, appear on the lower surfaces of leaves.
2. In other cases, only chlorotic areas ("oil flecks") are visible on the leaves. Additional symptoms may include narrow, folded leaves; bud abortion; twig dieback; or terminal tufts of leaves.
3. Fruits of black cherry develop red spots as the yields decline.
4. Necrotic symptoms appear on secondary roots, and cankerous wounds and gummosis develops on branches.

Control measures

The diseases are not important enough to merit any particular control measures.

Cherry Blossom Anomaly Diseases

Causal organism: X Disease Phytoplasma.

Symptoms

1. Leaf symptoms are similar to those of boron deficiency as the boron content of diseased tissues is low. Leaves are elongated and narrow and curve inward. This may be accompanied by interveinal chlorosis. In older infections, leaves originate from terminal buds and form rosettes at the tips of bare branches.
2. Normally white petals may exhibit varied amounts of green coloration starting at the midrib. This usually shows as stripes, but may encompass the entire petal. Some petals appear leaf like and some flower buds look like leaf buds, except for their yellow anthers.
3. Fruits developing from affected blossoms tend to be pointed, have roughened skin, and usually have a persistent style.

Control measures

The diseases are not important enough to merit any particular control measures.

Cherry Albino Disease

Causal organism: X Disease Phytoplasma.

Symptoms

Nearly the same symptoms appear in all sweet cherries, differences occur only in the sensitivities of each cultivar.

1. Leaves roll inward and turn a golden bronze or olive-brown in color as early as June.
2. Coloring appears mostly along the main or secondary veins to form feathery patterns.

3. A general symptom of the disease has the leaves produced in late summer from terminal buds forming rosettes.
4. Fruits remain small and fail to mature, assume a light color, and drop prematurely. In some cases, small white fruits remain on the tree longer.

Control measures

The diseases are not important enough to merit any particular control measures.

Apricot Chlorotic Leaf Roll Disease

Causal organism: European stone fruit yellows phytoplasma.

Symptoms

Symptoms of this disease vary; young trees may become systematically infected, whereas in trees older than 5 years, initial symptoms are localized to lower branches and spread to the crown. In either case, infected trees do not live long.

1. Early bud opening and foliation are reliable symptoms.
2. Characteristic symptoms first appear in May as the blades roll inward along convergent lines and turn pale and chlorotic.
3. Leaf fall is delayed, and abnormal bud opening continues in freezing temperatures.
4. Fruits are undersized, sometimes bumpy, and drop prematurely. The flesh is brown and spongy near the pit.
5. Necrosis may develop in the bark, and is visible in transverse sections as either thin orange lines or thicker brown bands.

Control measures

The diseases are not important enough to merit any particular control measures.

Another Mycoplasmal Diseases

There are following some another mycoplasmal diseases are also found in different plants.

(a) ***Witches' broom of Chinese jujube*** (*Ziziphus jujuba* Mill.) is found in China and Korea. The causal agent has been confirmed to be a mycoplasma-like organism. The MLO is transmitted by a leaf hopper, *Hishimonoides chinensis* Anufrive.

(b) ***Witches' broom of longan*** (*Euphoria longan* Lam.) is found in China and neighboring countries. A flexible virus has been isolated from infected longan trees. The virus is transmitted by a stinkbug,

Tessaratoma papillosa Drury, and a psylliid, Cornegenapsylla sinica Yang et Li. It is transmitted from one longan tree to another, and also from longan to litchi (*Litchi chinensis* Sonnerat). The symptoms in longan are very similar to those produced by litchi witches' broom disease.

(c) ***Marginal flavescence of potato disease*** is prevalent in hills and the plateau area where cool weather prevails for a longer time during the crop season. The loss due to this disease may vary from 40 to 75 per cent. The symptoms of Marginal flavescence of potato disease are chlorosis on the margins of upper leaves. In rare cases the severely affected plants, after exhibiting flavescence of leaf margins for quite some time develop pinkish colouration of leaflet bases and start rolling. Hairy root symptoms on tubers of 60 to 65 percent plants affected with MF have been observed. The pathogen is leafhopper and graft transmitted. Five to 35 per cent tubers carry the pathogen and serve as a source of primary inoculum. Hot water treatment of tubers at 500°C for 10 to 15 minutes eliminates the pathogen without adversely affecting the tubers.

(d) ***Withes broom of potato disease*** results in extreme stunting of the plant and numerous filamentous stems with simple leaves. The affected plants resulting in total loss of the plant produce very small tubers. The pathogen is leafhopper transmitted and perennates through tubers.

INDEX